AF412872

Radiofrequency in Cosmetic Dermatology

Aesthetic Dermatology

Vol. 2

Series Editor

David J. Goldberg New York, N.Y.

Radiofrequency in Cosmetic Dermatology

Volume Editors

Moshe Lapidoth Petach Tikva

Shlomit Halachmi Herzelia Pituach

46 figures, 36 in color, and 2 tables, 2015

Basel · Freiburg · Paris · London · New York · Chennai · New Delhi · Bangkok · Beijing · Shanghai · Tokyo · Kuala Lumpur · Singapore · Sydney

Dr. Moshe Lapidoth
Dermatologist, Head of Laser Unit
Dermatology Department
Rabin Medical Center
7 Keren Kayemet St.
Petach Tikva
Israel

Dr. Shlomit Halachmi
Herzelia Dermatology and Laser Center
9 Hamenofim St.
Herzelia Pituach
Israel

Library of Congress Cataloging-in-Publication Data

Radiofrequency in cosmetic dermatology / volume editors, Moshe Lapidoth, Shlomit Halachmi.
       p. ; cm. -- (Aesthetic dermatology, ISSN 2235-8609 ; vol. 2)
   Includes bibliographical references and indexes.
   ISBN 978-3-318-02316-9 (hard cover : alk. paper) -- ISBN 978-3-318-02317-6 (electronic version)
   I. Lapidoth, Moshe, editor. II. Halachmi, Shlomit, editor. III. Series:
Aesthetic dermatology (Series) ; v. 2. 2235-8609
   [DNLM: 1. Cosmetic Techniques. 2. Radio Waves--therapeutic use. 3. Skin Diseases--therapy. 4. Skin--pathology. WR 650]
   RL113
   616.5′0642--dc23
                                                                2014042585

Bibliographic Indices. This publication is listed in bibliographic services, including Current Contents®.

© Copyright 2015 by S. Karger AG, P.O. Box, CH-4009 Basel (Switzerland)
www.karger.com
Printed in Germany on acid-free and non-aging paper (ISO 9706) by Kraft Druck, Ettlingen
ISSN 2235–8609
e-ISSN 2235–8595
ISBN 978–3–318–02316–9
e-ISBN 978–3–318–02317–6

Contents

Preface

Radiofrequency (RF) current is an electrical current, typically with a frequency in the megahertz range. RF has been known to medicine since the late 19th century; however, the usage of RF in cosmetic and aesthetic medicine started only less than two decades ago.

The basic idea behind using RF on the skin is its ability to deliver heat to depth. Lasers and light sources are also able to generate heat, but their energy is absorbed in a chromophore-specific manner (the theory of 'selective photothermolysis'). RF electrical conductivity creates heat in the tissue based on the properties of the tissue itself (like the skin temperature and water content) and is not dependent on specific chromophores.

Today, RF current can be applied to the skin in various ways: unipolar or monopolar, bipolar, 'multipolar' (multiplication of bipolar) and combinations of RF and light, ultrasound, magnetic field, vacuum, etc., all of them commonly used for tissue heating and tightening.

Recently, after the introduction of fractional photothermolysis, which applies light-based techniques that enable the formation of an array of microscopic thermal ablated or coagulated wounds in the skin to induce a therapeutic healing response throughout the skin layers, several fractional RF devices have joined this armamentarium. These devices can be divided into superficial ablation systems and minimally invasive (microneedling RF) systems.

The aim of this book is to introduce to the reader the variety of RF techniques currently available on the market and to evaluate the efficacy and safety of typical RF devices in rejuvenating the skin, in improving the signs of facial aging, and providing skin tightening, to better understand and to support informed decisions in choosing the right treatment options for us and for our patients.

Moshe Lapidoth, Petach Tikva

Lapidoth M, Halachmi S (eds): Radiofrequency in Cosmetic Dermatology.
Aesthet Dermatol. Basel, Karger, 2015, vol 2, pp 1–22 (DOI: 10.1159/000362747)

Basic Radiofrequency: Physics and Safety and Application to Aesthetic Medicine

Diane Irvine Duncan[a] · Michael Kreindel[b]

[a]Plastic Surgical Associates of Fort Collins, Fort Collins, Colo., USA; [b]Invasix Corp., Richmond Hill, Ont., Canada

Abstract

This chapter summarizes the basic science of radiofrequency (RF) and its application in aesthetic medicine. The main parameters of RF including RF frequency, waveform, power, pulse duration, and penetration depth are described, and its application for treatment is analyzed. Monopolar and bipolar devices are described in detail for different clinical applications. The effect of RF electrode geometry on tissue heating is shown, and tissue-specific electrical parameters are summarized. The chapter discusses which RF parameters are required to reach therapeutic temperatures for tissue ablation, coagulation, or subnecrotic heating. RF parameters used for noninvasive, minimally invasive, and fractional treatment are compared. Finally, the chapter explains the main safety concerns associated with RF treatments and details the most common causes of adverse events.

The term radiofrequency (RF) was first introduced with the invention of radio and was applied to electromagnetic radiation or current ranging from 3 kHz to 300 GHz. Since then, the field of medicine has used the relatively narrow band of this spectrum from 200 kHz to 40 MHz in many different applications. The main advantage of RF energy in medicine is a low or negligible reaction of nerves to high-frequency alternating current (AC) in comparison to lower frequencies.

William T. Bovie invented the first electrosurgical device while working at Harvard [1]. This device was used by Dr. Harvey Williams Cushing on October 1, 1926, at Peter Bent Brigham Hospital in Boston, Mass., to remove a tissue mass from a patient's head [2]. Since then, RF electrosurgical devices have become one of the most useful surgical instruments. Recently, RF has experienced a resurgence in aesthetic medicine with applications for ablative and nonablative applications. RF energy has become an irreplaceable tool in almost every field of medicine including dermatology, plastic surgery, and aesthetic medicine, the primary interest of this book. The tissue effects achievable using RF energy are based on a versatile thermal end point and are dependent on the applied energy density.

Several RF-induced thermal changes of tissue are commonly used in medicine:

(1) Ablation of tissue. This effect is generally used for cutting or removing tissue and is based on thermal evaporation of tissue. Ablation requires very high energy density, allowing conversion of tissue from a solid state to vapor with minimal thermal damage to the surrounding tissue [3]. A new use for RF ablation is for cautery of tumors.

(2) Coagulation. When applied to blood vessels, coagulation provides hemostasis for controlling bleeding during surgery. The same mechanism is effective for vascular lesion treatment [27]. Coagulation may be applied to soft tissue as well, to induce necrosis when immediate tissue removal is not required or not practical.

(3) Collagen contraction. High temperatures induce immediate transformation in the tertiary structure of proteins. When applied to collagen, heating allows tissue shape to change for medical and cosmetic purposes. Immediate, predictable collagen contraction occurs at a temperature range of 60–80°C in orthopedic procedures [4] and ophthalmology [5]. For noninvasive cosmetic procedures, this effect is produced with lower temperatures in order to avoid skin necrosis. However, due to the lower temperatures, the outcome of the procedure is often less consistent, requires multiple procedures, and takes a longer time to show results [6, 7].

(4) Tissue hyperthermia. Heating of tissue to superphysiologic temperatures is a popular method of skin treatment using subnecrotic temperatures to stimulate natural physiological processes in attempts to modify skin appearance and to reduce subcutaneous fat [8, 9]. This heating does not induce immediate effects of coagulation but can stimulate fibroblasts to synthesize collagen and may alter the metabolism of adipocytes in favor of lipolysis.

Radiofrequency Energy Characteristics

The clinical effects of RF depend on a combination of the RF parameters and on the method of its application to the tissue.

Radiofrequency Frequency

The frequency of electrical current characterizes how many times per second an electrical current changes its direction and is reported in hertz. This change in direction is associated with a change of voltage polarity. Direct current has a frequency of 0 Hz, which is typically used in battery-powered devices. Standard AC in the range of 50–

Duncan · Kreindel

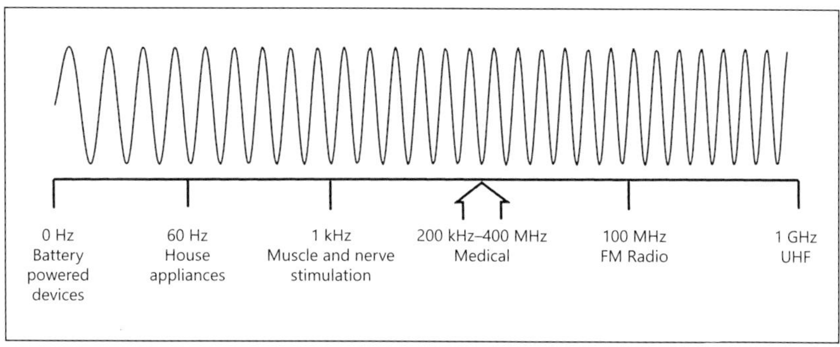

Fig. 1. Frequency spectrum of electrical current.

60 Hz is used for most home appliances. AC current causes nerve and muscle stimulation and at high powers is very dangerous. It can cause acute pain, muscle spasms, and even cardiac arrest.

At a frequency of 100 kHz and higher, the muscle- and nerve-stimulating effects decrease. In this range, higher power can be applied to the tissue safely to create the desired thermal effect (fig. 1). Although at frequencies above 100 Hz nerve reaction from electrical current is dramatically diminished, at high amplitudes skin reaction can be observed even at a frequency above 1 MHz. RF energy propagates in the tissue in the form of electrical current between applied electrodes and in the form of radiation at higher frequencies. Frequencies in the range of 200 kHz to 6 MHz are the most common in medicine, but there are devices with frequencies up to 40 MHz [10]. The higher frequency electrical oscillations are used mostly for communication.

Radiofrequency Waveform

Typically, sine RF voltage is used in medical devices. The RF energy can be delivered in continuous wave (CW) mode, burst mode and pulsed mode (fig. 2). For gradual treatment of large areas, the CW mode is most useful as it allows a slow increase in temperature in bulk tissue. This approach is applied for targeting cellulite, subcutaneous fat, and skin tightening. The burst mode delivers RF energy with repetitive pulses of RF energy. It is used in applications where peak power is important while average power should be limited. This application is used in blood vessel coagulation. Pulsed mode is optimal when the goal is to heat a small tissue volume while limiting heat conduction to the surrounding tissue, similar to the rationale of applying short pulse duration in laser treatments. Pulsed mode is effective for fractional skin ablation and is characterized by pulse durations which do not exceed the thermal relaxation time (TRT) of treated zone.

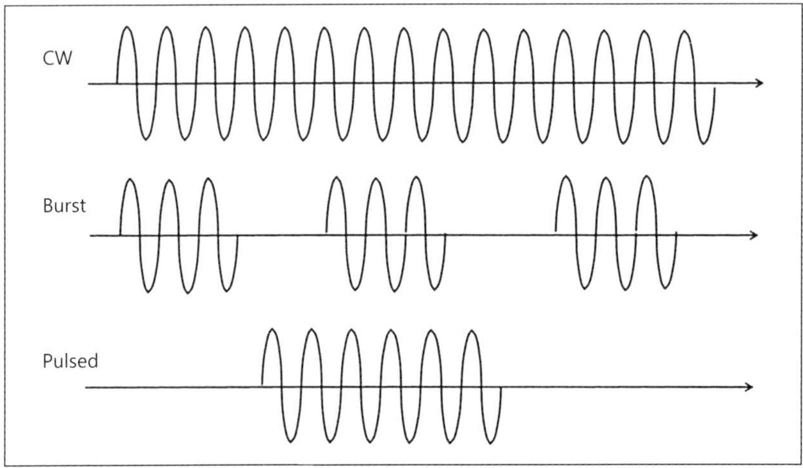

Fig. 2. Typical RF waveforms.

Radiofrequency Power

The most important characteristics of RF energy are its peak and average power. Peak power is important to estimate the thermal effect produced, while average power affects the speed at which the heating is induced. For CW operation mode, the peak and average power are the same. For pulsed or burst mode, the average power is the total power delivered divided by the time the device is applied, including the 'off' cycles.

Another important characteristic of RF is power density. High power applied to a large skin surface may create only gentle warming, but when applied through a needle electrode, the same power is applied over a small contact point, leading to high power density. At high power densities, RF may create intense tissue ablation rather than warming or coagulation.

Thermal Effect of Radiofrequency Current

The heat power (P) generated in a tissue volume by electrical current during a period of time (t) is described by Joule's law:

$$P = \frac{j^2}{\sigma}. \tag{1}$$

The heat generated is measured in joules/cm^3. As the equation describes, power increases as a square function of the RF current density (j). Conversely, heating power changes in inverse proportion to tissue conductivity (σ).

Duncan · Kreindel

Taking into account that current density according to Ohm's law is proportional to the electric field strength and tissue conductivity (equation 2),

$$j = \sigma E \tag{2},$$

we can rewrite the equation (1) as

$$P = \sigma E^2 \tag{3},$$

In other words, the higher the tissue conductivity, the greater the heat that will be generated when constant RF voltage is applied between the electrodes. In addition, the amount of heat generated increases with increasing exposure to RF; stated differently, tissue will heat more with longer duration of RF current. As tissue heats, its conductivity increases (or, stated differently, impedance decreases), and the equations are therefore relevant only at a given time. This is taken into account during RF procedures: in modern devices, RF power is automatically adjusted to tissue impedance.

Penetration Depth and Radiofrequency Energy Distribution Between Electrodes

Penetration depth is a parameter broadly used in laser dermatology to mean the distance below the skin which is heated. More correctly, the depth of RF effect is characterized by attenuation of applied energy with the depth. The most common understanding of this parameter is a depth where applied energy is decreased by an exponential factor (e ~2.7). In contrast to optical energy, which is attenuated with distance of travel through tissue as a result of scattering and absorption, RF current decreases at a distance from the electrode due to the divergence of current lines. The depth of penetration can be affected by altering the topology of the skin and optimizing the electrode system. In aesthetic medicine, the most common configurations of electrode systems are monopolar, bipolar, and multipolar including fractional, where the effect is achieved by superposition of RF current paths between paired electrodes. Penetration depth also can be affected by the anatomical structure of treated area. For example, penetration depth over a bone can be limited by low conductivity of bone tissue. For this reason, treatment parameters over bone, for example the forehead and hip, often differ from the parameters applied in adjacent areas.

Monopolar Radiofrequency Systems

Monopolar RF devices utilize an active electrode in the treatment area and a return electrode, usually in the form of a grounding pad with a large contact area, which is placed outside of the treatment zone. In this electrode geometry, a high RF current

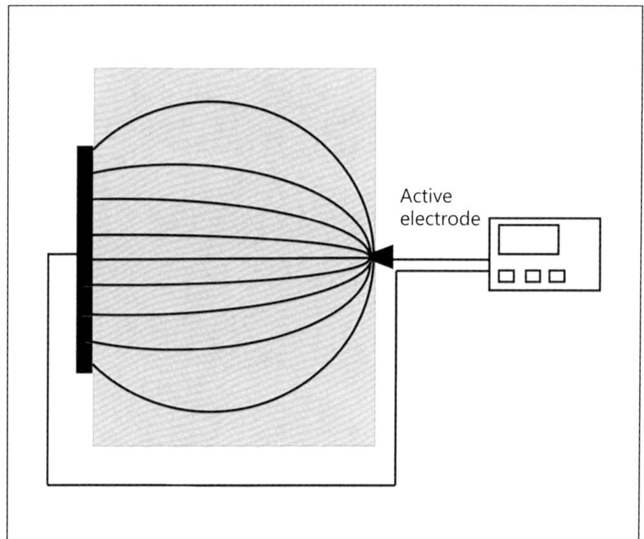

Fig. 3. Schematic of RF current distribution between electrodes for a monopolar system.

density is created near the active electrode, and the RF current diverges toward the large return electrode. Schematically, RF current behavior in the body for a monopolar system is depicted in figure 3.

The heat zone for this geometry can be estimated using an analytic spherical model for the continuity equation, stating that electrical current flows continuously from one electrode to another:

$$\nabla_r j = 0 \tag{4}$$

Taking into account Ohm's law in differential form (equation 2) and the definition of an electric field, equation 4 can be rewritten as:

$$\frac{1}{r^2}\frac{\partial}{\partial r} r^2 \frac{\partial \varphi}{\partial r} = 0, \tag{5}$$

where φ is the potential of the electric field. The solution for this equation provides the RF current density distribution between electrodes:

$$j = \frac{\sigma V r_0 R}{r^2 \left(R - r_0\right)}, \tag{6}$$

where σ is tissue conductivity, V is voltage between electrodes, r_0 is radius of small electrode and R is the radius of the large electrode.

For the instance when the return electrode is much larger than the active electrode, the equation can be simplified as:

$$j = \frac{\sigma V r_0}{r^2}. \tag{7}$$

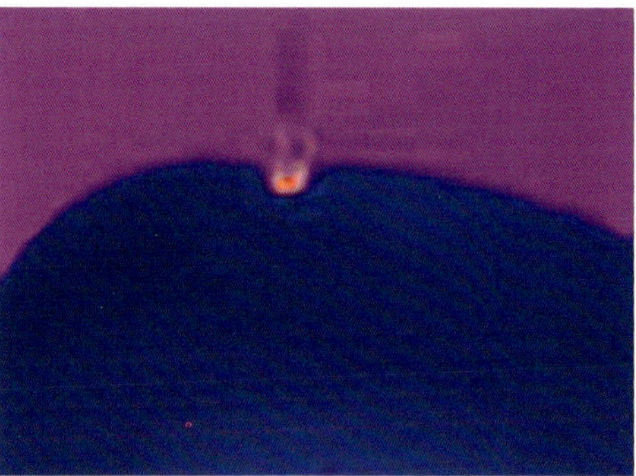

Fig. 4. Thermography of tissue in cross-section during treatment: a monopolar RF generator with a frequency of 1 MHz and 50-watt power was applied using a 1-mm electrode at the tissue surface and a large 100-cm^2 return electrode at the bottom of the tissue. The heat is concentrated near the surface of the small electrode, and the depth of thermal zone is half of the electrode size.

Correspondently, heat power according to Joule's law can be estimated as:

$$P = \frac{\sigma V^2 r_0^2}{r^4} \tag{8}$$

This simple equation leads to a few interesting conclusions:

(1) Heat generated by RF current near the active electrode does not depend on the size, shape, or position of the return electrode when the return electrode is much larger in size than the active electrode and is located at a distance which is much greater than the size of the active electrode.

(2) Heating decreases dramatically as distance increases from the electrode. At a distance equal to the electrode size, heating becomes insignificant. In other words, most of the RF energy applied in monopolar systems is converted into heat near the active electrode. Therefore, the heat zone can be estimated as a radius or half size of active electrode.

(3) RF current is concentrated on the RF electrode and rapidly diverges toward the return electrode. Figure 4 shows a thermal image in cross-section of bovine tissue treated with a monopolar electrode and demonstrates that heat generation is observed near the active electrode only.

Monopolar devices are most commonly used for tissue cutting. Schematically, the RF current flow for monopolar devices is shown in figure 5.

RF current always flows in a closed loop via the human body. As shown above, the current density far from the active electrode is negligible. However, a malfunction in which low frequency current escapes from a monopolar configuration holds high risk because the entire body is exposed to the electrical energy. Most commercially available devices have isolated output to help avoid any unexpected RF current path to the surrounding metal equipment.

Treatment effects with monopolar devices depend on the density of RF energy, which can be controlled with RF power, and the size of active electrode. In order to create tissue ablation, very high energy density is required. In cutting instruments, a needle type electrode is used to concentrate electrical current on a very small area.

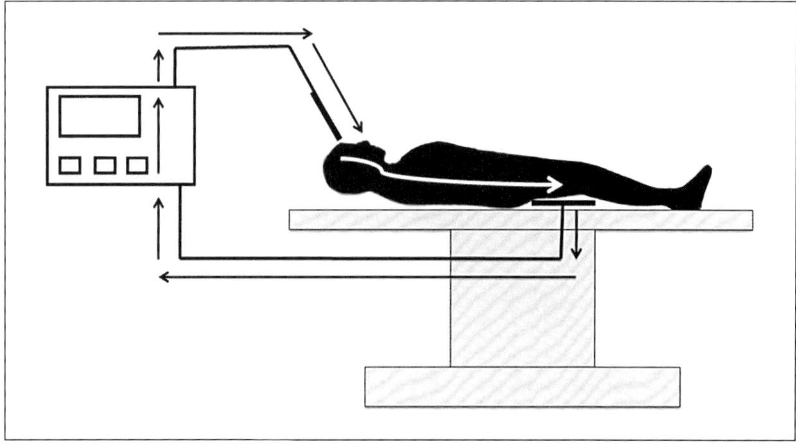

Fig. 5. Electrical current flowing through the patient and electrosurgical device.

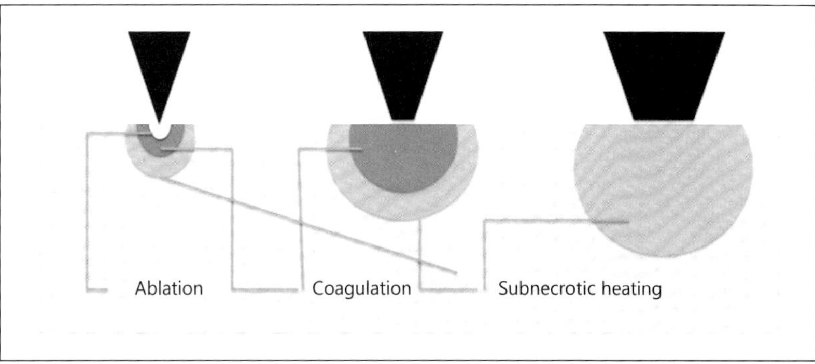

Fig. 6. Spot size effect.

Coagulation hand pieces have a larger surface area than ablative devices, usually a few square millimeters, to generate heat on a larger area, creating coagulation rather than ablation. Subnecrotic heating is usually used for treatments related to collagen remodeling, and in this case the spot size is about 1 cm^2 [7]. A schematic of the spot size effect on the treatment area is shown in figure 6.

For monopolar devices, the penetration depth is a function of the active electrode size and can be estimated as a half the electrode size.

The main features of monopolar devices are:
- Predictability of thermal effect near the active electrode
- Ability to concentrate energy on a very small area
- High nonuniformity of heat distribution, with very high heat at the surface of the active electrode and dramatic reduction at a distance exceeding the size of electrode, thereby limiting penetration depth.

Duncan · Kreindel

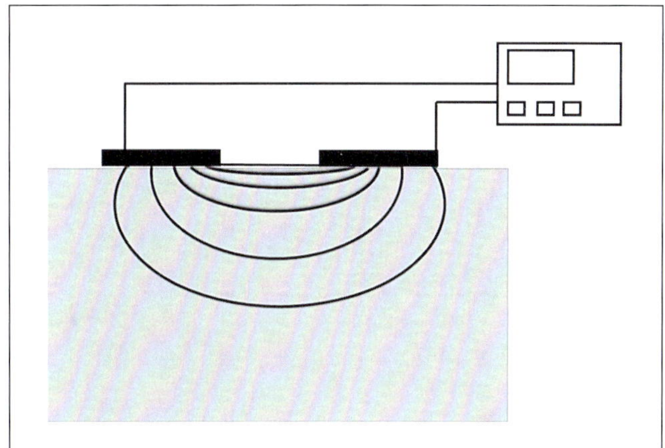

Fig. 7. Electrical current distribution for a bipolar RF system.

Bipolar Radiofrequency Systems

Bipolar configuration is characterized by the use of two electrodes which are in contact with the treated area. This geometry is better able to create uniform heating in larger volume of tissue than a monopolar system. In order to understand heat distribution between electrodes, the following three rules should be taken into the account:

(1) For any geometry, RF current density is higher along the line of shortest distance between the electrodes and reduced with distance from the electrodes.

(2) Heating is greater near the electrode surface and drops with distance because of current divergence.

(3) RF current is concentrated on the part of the electrode that has high curvature, creating hot spots.

A schematic distribution of electrical currents in uniform media for typical electrode geometries used for noninvasive treatment is shown in figure 7.

In bipolar devices, both electrodes create an equal thermal effect near each of the electrodes, and the divergence of RF current is not strong because of the small distance between the electrodes. For bipolar systems shown in figure 7, most of the heat is concentrated between electrodes.

Penetration depth of RF for bipolar devices is a function of electrode size and the distance between them. By increasing the distance between the electrodes, electrical current can go deeper, but divergence is also increased. For the case when the distance between the electrodes is much larger than the electrode size, the heating profile will be similar to two monopolar electrodes. Schematically this situation is shown in figure 8.

Thermal images of tissue cross-section for small and large distance between electrodes are shown in figure 9.

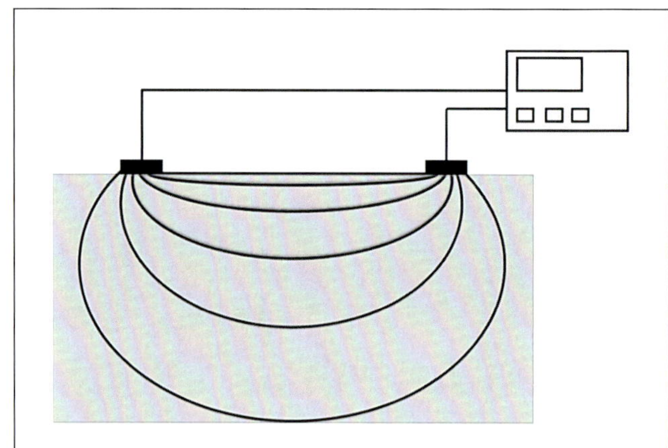

Fig. 8. Electrical current distribution for a bipolar system with a large distance between electrodes.

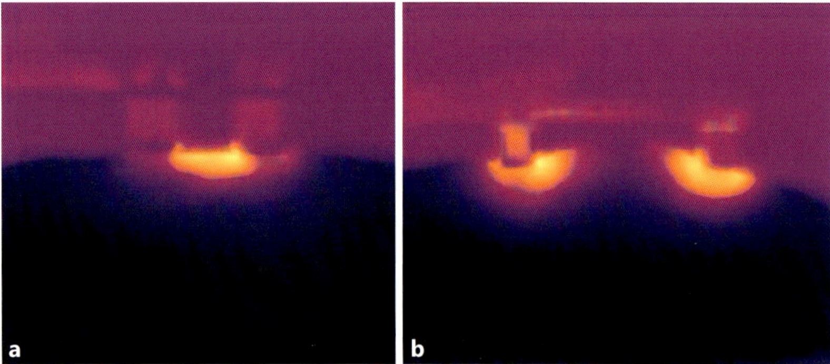

Fig. 9. Thermal images of tissue cross section treated with bipolar device at small (**a**) and large (**b**) distance between electrodes.

In figure 9a, the heat is generated between the electrodes, while the heating profiles directly under the electrodes are less pronounced. This geometry allows generation of uniform heat in a limited volume. This geometry is suitable for homogeneous heating of the skin layer with a depth of up to a few millimeters. The main application of this geometry is subnecrotic skin heating for collagen denaturation and stimulation of re-modeling. In figure 9b, the heat is concentrated under the electrode, as occurs in monopolar devices. The temperature distribution is not uniform, and in practice it is evident the heating occurs with hot spots.

The most uniform distribution of RF current is obtained in planar geometry when the area of parallel electrodes is larger than the distance between them. RF current distribution for planar geometry is shown in figure 10.

RF heating between electrodes will be uniform for most of the volume with divergence of current at the periphery of the electrodes. This geometry can be reached by

Duncan · Kreindel

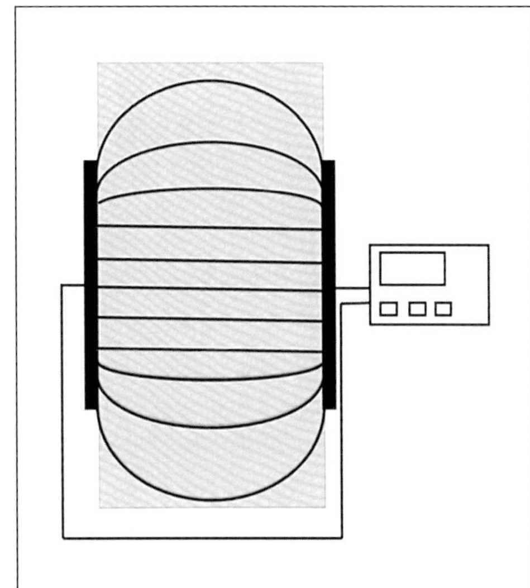

Fig. 10. Electrical current distribution for planar bipolar geometry.

folding tissue between electrodes. This is commonly done in aesthetic medicine by applying negative pressure (in the form of vacuum) to elevate and pinch the skin between two parallel electrodes. This geometry is typically used in body contouring to deliver uniform heating to depth.

Bipolar devices are usually used to create larger thermal zones in nonablative applications. The advantage of bipolar systems is the localization of electrical current in the treatment area.

The response of tissue to bipolar RF can be demonstrated by thermal experiments conducted in in vitro studies using porcine tissue. For the current example, an RF generator with a frequency of 1 MHz and 50-watt power was applied. A thermal camera (FLIR A320) was used for thermography of tissue during RF application. Figure 4, earlier in this chapter, shows the thermal response to monopolar RF where a 1-mm electrode was applied to the tissue surface and a large, 100-cm^2 return electrode was placed at the bottom of the tissue. The heat is concentrated near the surface of the small electrode, and the depth of thermal zone is about half of the electrode size. In contrast, figure 9b shows bipolar geometry where both electrodes have an equal size of 10 mm and the distance between them is 10 mm. The thermal zone is located between electrodes and has uniform distribution down to a depth of 5 mm. For bipolar geometry, where the distance between the electrodes is about electrode size or less, the penetration depth is about half of the distance between electrodes. At an increasing distance between the electrodes, the RF energy distribution becomes nonuniform, and most of the heat is concentrated near the electrode surface (fig. 9b). Folding the skin between two planar electrodes allows uniform heating of large tissue volume (fig. 11). Penetration depth is determined by electrode height and can be as large as a few centimeters.

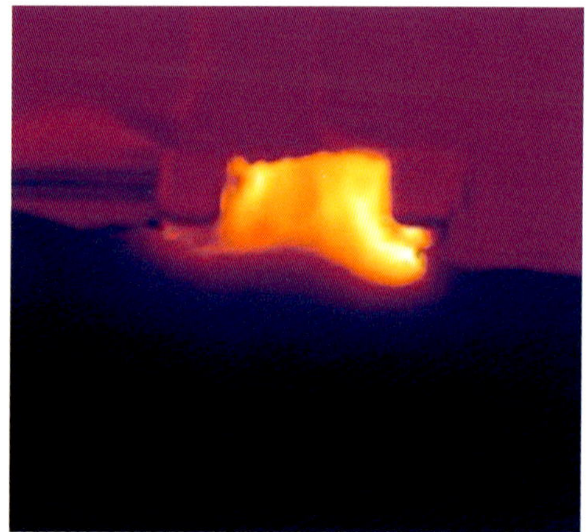

Fig. 11. Thermal image of heat distribution created in the skin folded between two parallel electrodes.

Electrical Properties of Tissue

A specific feature of RF current in biological tissue is ion conductivity. As a result, the electrical effects related to magnetism are negligible, and tissue behavior under RF current is quite well described using Maxwell theory. Considering tissue as a resistant media having some capacitive properties, this has an effect which becomes more significant at higher frequencies. In the RF range of 200 kHz to 1 MHz, the tissue resistivity significantly dominates in tissue behavior, and we can ignore capacitive properties, which are more significant for RF generator development than for medical applications. Therefore, for purposes of this discussion, the terms resistance and impedance will be considered the same.

For tissue with uniform properties, resistance (R) is equal to:

$$R = \rho \frac{L}{S}, \tag{9}$$

where ρ is resistivity of tissue, which is equal to resistance of a conductor with an area of 1 m^2 and length of 1 m. S is the cross-section of tissue experiencing RF current and L is the distance between electrodes. This simplified equation allows comprehension of the most basic principles of RF current behavior: tissue impedance is higher for smaller electrodes and a larger distance between them.

Often in literature, the term conductivity is used as the opposite to resistivity. Conductivity of different types of tissue may vary significantly. Electrical properties of some tissues are presented in table 1.

It is critical to understand that in vitro measurements for pure substances can be significantly different from a living patient because on a macro level there is a mix of tissues. For example, according to the table above the difference between wet skin and fat is approximately a factor of 8, while at multiple measurements conducted in vivo

Table 1. Conductivity of different types of biological tissue at 1 MHz [12]

Tissue	Conductivity, $S\,m^{-1}$
Blood	0.7
Bone	0.02
Fat	0.03
Dry skin	0.03
Wet skin	0.25

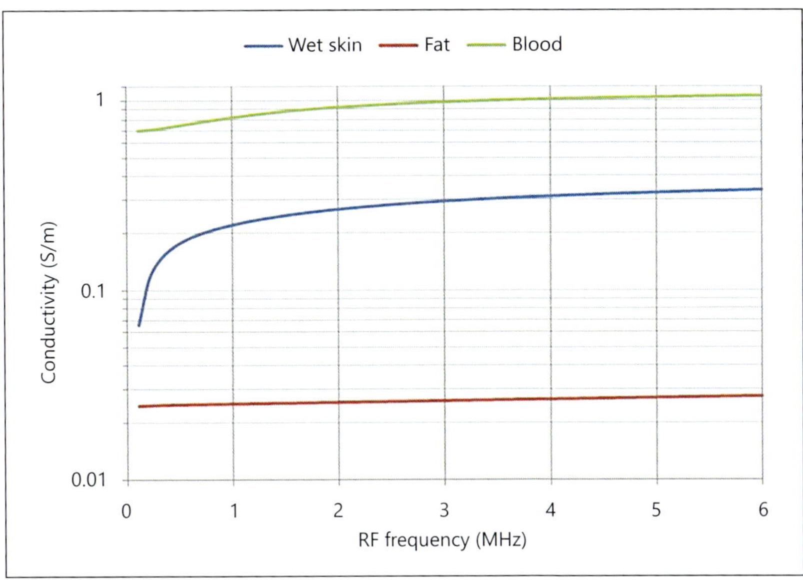

Fig. 12. Tissue conductivity as a function of RF frequency [12].

the difference is approximately a factor of 3. This can be explained by the presence of a vascular network, connective tissue matrix, and intercellular liquids in the adipose layer. It can also explain the significant variance in data reported in different studies [11]. Basically, tissue with higher water and blood content has high electrical conductivity. Tumescent anesthesia may significantly increase tissue conductivity by increasing water and salt content.

Tissue conductivity can be a strong function of RF frequency. Figure 12 shows conductivity of fat and skin calculated according to the parametric model [12]. Skin conductivity is strong function of frequency in the range of 100 KHz to 1 MHz and has a weak change at higher frequencies. Fat conductivity is flat in all the ranges of frequencies used in medicine.

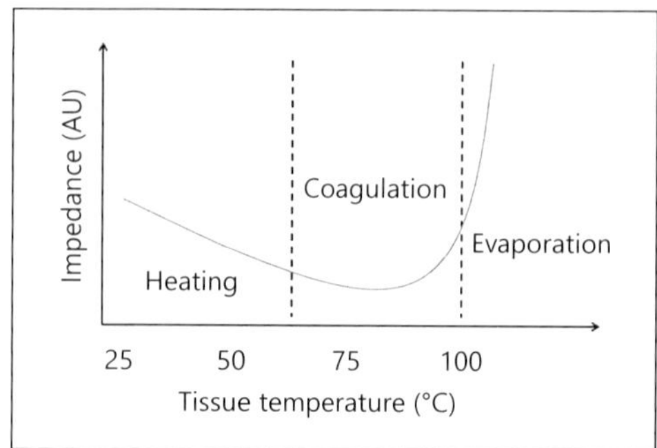

Fig. 13. Tissue conductivity as a function of temperature.

Tissue electrical conductivity is a function of temperature. Qualitative behavior of tissue impedance as the function of temperature is shown in figure 13.

Warming of tissue reduces its impedance with a rate of about 1.5–2% per degree centigrade up to the point of coagulation [11]. This change is related to reduction of tissue viscosity which is reduced with temperature increase. Coagulation of the tissue causes a chemical change in tissue structure, and the trend of impedance behavior is changed. When tissue is heated to 90–100°C, the evaporation of liquids starts, which increases tissue impedance substantially. Further heating of tissue leads to its carbonization. The dependence of tissue conductivity on temperature is utilized by some medical devices. For example, a technology known as electro-optical synergy applies light in particular wavelengths for preferential heating of certain tissue targets; the preheating of the target tissue then creates a preferable path for RF current [13, 14]. This can provide treatment advantages for some applications.

Radiofrequency Thermal Effect on Tissue

The thermal effect of RF on tissue is not different from laser or any other heating method. Multiple studies [15, 16] discuss the temperature effect on tissue. Treatment effect is not a function of temperature only, but also of the length of time when this temperature is applied. Therefore, exposure to a temperature of 70–90°C for milliseconds can cause coagulation, while temperature applied for a few seconds at a lower temperature of 45°C causes irreversible damage.

The typical sequence of tissue response to temperature increase is as follows. 37–44°C: acceleration of metabolism and other natural processes. 44–45°C: conformational changes in proteins, including collagen; hyperthermic cell death. 60–70°C: denaturation of proteins; coagulation of collagen, membranes, hemoglobin; shrinkage of collagen fibers. 90–100°C: formation of extracellular vacuoles; evaporation of liquids. >100°C: thermal ablation; carbonization.

Pulse Duration Effect

Pulse duration is one of the most critical parameters when utilizing RF energy in order to achieve a clinical response. It affects treatment results because timing influences the thermochemical process in tissue. The other effect of pulse duration is energy dissipation away from the treatment zone due to heat conductivity from the exposed area to the surrounding tissue.

There is extensive data on the correlation between tissue temperature, pulse duration, and treatment effect. Moritz and Henriques [17] demonstrated that the skin thermal damage threshold is a function of temperature and time. Later, it was demonstrated that skin damage function can be described by the Arrhenius equation, where time is a preexponential factor and temperature is an exponential factor [16]:

$$D = A\, t\, \exp\left(\frac{-\Delta E}{R\, T}\right).$$

In other words, the degree of damage *(D)* is a linear function of pulse duration *(t)* and an exponential factor of tissue temperature *(T)*. Practically speaking, then, tissue temperature is more influential on the degree of damage than pulse duration. Nonetheless, prolonged low-grade temperature elevation impacts tissues [27].

It is well known that sustained hyperthermia at 42°C for tens of minutes causes death of most sensitive cells [18]. Once elevated, tissue temperature can only be reduced by dissipation of heat. Temperature dissipation is characterized by the TRT of the targeted tissue. When treatment is intended to heat a structure without heating the peripheral tissue, it must be elevated to that temperature before dissipation begins by heat transfer. Therefore, to localize treatment, the pulse duration should be less than the TRT.

The TRT is a function of tissue thermal properties, as well as the shape and size of the heated volume. Soft tissue has thermal properties close to water.

For a planar object, the TRT can be estimated as [19]:

$$TRT = \frac{d^2}{4\, a},$$

where d is the thickness of the layer and a is tissue diffusivity. Diffusivity is equal to tissue conductivity divided by the heat capacitance and is measured in $cm^2\, s^{-1}$.

For a cylindrical object, such as a blood vessel or hair, a similar equation can be used with different geometrical factors:

$$TRT = \frac{d^2}{16\, a},$$

where d is object diameter. The equation makes evident that cooling time is a square function of the size of the heated target [26].

Radiofrequency Applications

In aesthetic medicine, the RF applications can be divided into three main groups:
- Noninvasive tissue heating with RF, which is used in a range of clinical applications including wrinkle reduction, skin tightening, cellulite, and circumference reduction
- Fractional coagulation and ablation for skin resurfacing
- Minimally invasive treatment for volumetric collagen shrinkage and fat melting.

Noninvasive Radiofrequency

Noninvasive RF treatment is based on the application of RF electrodes externally to the skin of the treatment area. The applied RF energy penetrates into the tissue up to a few millimeters. In order to reach collagenous tissue in the dermis and subcutaneous fat, the RF current must pass through the epidermis. There are some limitations to the amount of RF energy that can be applied noninvasively because the epidermal layer should remain undamaged. The limited heating results in a relatively conservative thermal effect, and usually multiple treatments are required to provide visible improvement. The RF energy can be applied using monopolar [7, 8] electrode geometry or bipolar systems. The RF energy can be delivered in pulsed mode, where a predetermined amount of energy is delivered to each spot, or in CW mode, in which electrodes move over the skin surface continuously for gradual, incremental heating. Typically, the temperature of the tissue should not exceed 40–43°C to avoid epidermal damage. Because skin damage is an exponential function of the temperature, it is challenging to get to the maximal point of the temperature range without the risk of a burn. It is much easier – and safer – to obtain optimal results by extending the treatment time and maintaining a safe temperature longer. The treatment effect is based mostly on collagen remodeling and local metabolism acceleration. Skin tightening, which is often desired in noninvasive treatments, requires heating of the reticular dermis and subdermal structures. The required heating depth for these indications is 3–6 mm, a range that light energy does not reach well; therefore, RF is currently the main tool for these kind of treatments [26]. For the indications of temporary improvement in the appearance of cellulite or circumference reduction, heating must be deeper. Vacuum can be used to assist in folding skin between electrodes and thereby to increase the penetration depth [20, 21].

Fractional Treatment

Fractional skin treatment was introduced in aesthetic medicine about a decade ago and has become one of the most popular modalities for the improvement of skin quality. This procedure is based on heating or ablation of multiple small foci with a spot size of 100–400 µm. This allows the procedure to be very tolerable and with relatively short downtime.

In contrast to lasers where the thermal effect is limited to the periphery of the ablation crater, RF energy flows through the whole dermis, adding volumetric heating to fractional treatment. This volumetric heating adds a skin-tightening effect. RF fractional technologies can be administered from the surface, using a grid of electrodes, or intradermally, using a grid of microneedles which deliver the RF energy within the dermis. The surface electrodes provide a more superficial effect improving texture and fine lines [19] while longer needles penetrate deeper, providing deeper dermal remodeling [22]. These approaches are described further in other chapters.

Minimally Invasive Radiofrequency Treatment

Minimally invasive RF treatment recently has gained popularity based on the patient's desire to obtain a more dramatic treatment result after a single treatment. Microneedle RF treats the skin in a minimally invasive manner. Dielectric coated needles have become popular in delivering aggressive heating to the reticular dermis without thermal damage to the skin's surface [23]. By heating deep dermal collagen at a higher temperature than could be safely used at the epidermal level, a much stronger collagen contraction effect can be achieved in order to improve deep wrinkles and enhance skin tightening. The combination of deep dermal treatment with superficial fractional treatment has a high potential for complete skin improvement while avoiding skin excision.

By introducing larger needle electrodes into the deep dermis, for example in RF-assisted liposuction, RF can be used to address tightening of the fibroseptal network of the adipose layer with subsequent accommodation of the overlying skin during local fat removal. When energy is applied under the skin, the dermis and epidermis are relatively protected. More aggressive heating up to 60–70°C can be applied during treatment, creating immediate and more pronounced collagen contraction. In some clinical studies [24, 25] up to 42% area skin contraction was achieved after RF-assisted lipolysis.

Safety Features of Radiofrequency Technology

RF treatment is based on a thermal effect created in a treatment zone, and therefore the typical side effects associated with RF energy have thermal character. Most are related to overtreatment and nonuniformity of the thermal effect. Hot spots are an inherent problem of RF technology. Density of RF current is always higher on the

However, a guiding principal is that thinner tissue should be treated with lower power. In addition, lower maximum temperature is mandated when treating thin skin and soft tissue, such as the neck and face.

(3) Patient sensitivity varies significantly. Some patients are more sensitive to treatment than others, and we cannot always recognize which patients are more sensitive prior to treatment. Applying test pulses and adjusting based on patient preference can assist in determining the ideal setting for a given patient.

General Safety Approach Using Radiofrequency Technology

There are a number of methods to minimize the risk of adverse effects without compromising treatment efficacy. The following are the main methods that are applicable to almost all RF treatments.

(1) Use test spots in less visible areas to determine how the skin will react to treatment.

(2) Begin with lower settings and gradually increase energy to optimal/advanced parameters

(3) Use lower settings on:

(a) Small zones

(b) Bone prominences

(c) Areas with high curvature

(4) Always observe the immediate skin reaction

(5) Stop energy and treatment when there is any indication for concern and reassess continuation of treatment

(6) Do not rush treatment

The use of test pulses is a common technique in laser and RF medicine to test treatment parameters in a less visible area in order to identify optimal settings for the full RF treatment. It is important to observe the skin reaction after each test pulse and adjust parameters if required. Adverse events may not appear immediately; therefore, it may take a few minutes or even a day after pulsing for the full response to be visible. Even for patients treated previously with higher parameters, each new session should start with slightly lower settings, as skin reaction may be different due to seasonal skin dryness or recent exposure to sun.

Parameters should be adjusted according to the treatment area. When treating small zones, the applicator overlaps the same spot more often and the average RF energy applied is higher. In order to compensate for this effect, lower RF power settings are recommended. When treating over bony areas such as the forehead, RF energy application to the thin layer of tissue results in stronger heating. Reduction of RF power improves comfort and provides a greater level of safety for the patient. In addition, it is more difficult to keep electrodes in full contact with the tissue over bony and highly curved areas. Poor contact results in high RF energy density in the areas of contact, which generates hot spots and can cause patient discomfort and burns. For such areas, it is always recommended that the operator reduce RF power and use more gel or other coupling liquid. In addition, RF should be stopped when there is a change in the patient position, while pausing to observe skin reaction, while adding more gel, and so on. For safety, it is more important to learn how to stop the device than how to activate it.

As discussed, before an adverse skin reaction appears, there are warning signs that can be a signal when problems are minor. If left unattended to or ignored, these can result in more significant issues. By closely observing the skin as well as safety feedback data from the equipment, an operator can predict the skin's response to treatment and prevent or curtail thermal injury.

In general, all these recommendations can be summarized to one basic preface: the best device is highly dependent on the operator. Nothing is more supreme than one's own educated observation. The manufacturer's treatment recommendations reflect the average treatment pattern, but each patient is unique. It will take time to get comfortable with the technology, so it is important not to rush during the procedure. The time lost with a slower treatment can never be compared with time spent on the treatment of adverse effects and patient dissatisfaction.

Conclusions

RF-assisted medical devices have evolved dramatically within the last two decades. What used to be a simple array of fairly basic tools has now become an extremely sophisticated and sometimes confusing collection of options. There is quite a bit of value in understanding the way RF energy works. The information in this chapter can help a potential buyer of new equipment make a rational choice, based on goals of treatment and physics of the RF device in question. Even more importantly, the physician's understanding of his or her devices can maximize treatment outcomes and can minimize unwanted adverse events and complications.

References

1 O'Connor JL, Bloom DA, William T: Bovie and electrosurgery. Surgery 1996;119:390–396.
2 Cushing H: Electrosurgery as an aid to the removal of intracranial tumors: with a preliminary note on a new surgical-current generator by W.T. Bovie. Surg Gynecol Obstet 1928;47:751–784.
3 Hainer BL: 'Fundamentals of electrosurgery'. J Am Board Fam Pract 1991;4:419–426.
4 Obrzut SL, Hecht P, Hayashi K, Fanton GS, Thabit G III, Markel MD: The effect of radiofrequency energy on the length and temperature properties of the glenohumeral joint capsule. Arthroscopy 1998;14:395–400.
5 Babilas P, Shafirstein G, Bäumler W, Baier J, Landthaler M, Szeimies R-M, Abels C: Selective photothermolysis of blood vessels following flashlamp-pumped pulsed dye laser irradiation: in vivo results and mathematical modelling are in agreement. J Invest Dermatol 2005;125:343–352.

6 Asbell P, Maloney RK, Davidorf J, Hersh P, McDonald M, Manche E: Conductive keratoplasty for the correction of hyperopia. Trans Am Ophthalmol Soc 2001;99:79–87.

7 Weiss RA, Weiss MA, Munavalli G, Beasley KL: Monopolar radiofrequency facial tightening: a retrospective analysis of efficacy and safety in over 600 treatments. J Drugs Dermatol 2006;5:707–712.

8 Fitzpatrick R, Geronemus R, Goldberg D, Kaminer M, Kilmer S, Ruiz-Esparza J: Multicenter study of noninvasive radiofrequency for periorbital tissue tightening. Lasers Surg Med 2003;33:232–242.

9 Anolik R, Chapas AM, Brightman LA, Geronemus RG: Radiofrequency devices for body shaping: a review and study of 12 patients. Semin Cutan Med Surg 2009;28:236–243.

10 Nootheti PK, Magpantay A, Yosowitz G, Calderon S, Goldman MP: A single center, randomized, comparative, prospective clinical study to determine the efficacy of the VelaSmooth system versus the Triactive system for the treatment of cellulite. Lasers Surg Med 2006;38:908–912.

11 Duck FA: Physical Properties of Tissue. London, Academic Press Limited, 1990.

12 Gabriel S, Lau RW, Gabriel C: The dielectric properties of biological tissues. III. Parametric models for dielectric spectrum of tissues. Phys Med Biol 1996; 41:2271–2293.

13 Waldman A, Kreindel M: New technology in aesthetic medicine: ELOS electro optical synergy. J Cosmet Laser Ther 2003;5:204–206.

14 Sadick NS, Alexiades-Armenakas M, Bitter P Jr, Hruza G, Mulholland RS: Enhanced full-face skin rejuvenation using synchronous intense pulsed optical and conducted bipolar radiofrequency energy (ELOS): introducing selective radiophotothermolysis. J Drugs Dermatol 2005;4:181–186.

15 Thomsen S: Pathologic analysis of photothermal and photomechanical effects of laser-tissue interactions. Photochem Photobiol 1991;53:825–835.

16 Katzir A: Lasers and Optical Fibers in Medicine. San Diego, Academic Press, 1993.

17 Moritz R, Henriques FC Jr: Studies of thermal injury. II. The relative importance of time and surface temperature in the causation of cutaneous burns. Am J Pathol 1947;23:695–720.

18 Moringlane JR, Koch R, Schäfer H, Ostertag ChB: Experimental radiofrequency (RF) coagulation with computer-based on line monitoring of temperature and power. Acta Neurochir 1989;96:126–131.

19 Man J, Goldberg DJ: Safety and efficacy of fractional bipolar radiofrequency treatment in Fitzpatrick skin types V–VI. J Cosmet Laser Ther 2012;14:179–183.

20 Belenky I, Margulis A, Elman M, Bar-Yosef U, Paun SD: Exploring channeling optimized radiofrequency energy: a review of radiofrequency history and applications in esthetic fields. Adv Ther 2012;29:249–266.

21 Brightman L, Weiss E, Chapas AM, Karen J, Hale E, Bernstein L, Geronemus RG: Improvement in arm and post-partum abdominal and flank subcutaneous fat deposits and skin laxity using a bipolar radiofrequency, infrared, vacuum and mechanical massage device. Lasers Surg Med 2009;41:791–798.

22 Mulholland RS, Ahn DH, Kreindel M, Malcolm P: Fractional ablative radio-frequency resurfacing in Asian and Caucasian skin: a novel method for deep radiofrequency fractional skin rejuvenation. J Cosmet Dermatol Sci Appl 2012;2:144–150.

23 Willey A, Kilmer S, Newman J, Renton B, Hantash B, Krishna S, Mcgill S, Berube D: Elastometry and clinical results after bipolar radiofrequency treatment of skin. Dermatol Surg 2010;36:877–884.

24 Paul M, Blugerman G, Kreindel M, Mulholland RS: Three-dimensional radiofrequency tissue tightening: a proposed mechanism and applications for body contouring. Aesthetic Plast Surg 2011;35:87–95.

25 Duncan D: Improving outcomes in upper arm liposuction: adding radiofrequency-assisted liposuction to induce skin contraction. Aesthet Surg J 2012;32:84–95.

26 Emilia del Pino M, Rosado RH, Azuela A, Graciela Guzmán M, Argüelles D, Rodríguez C, Rosado GM: Effect of controlled volumetric tissue heating with radiofrequency on cellulite and the subcutaneous tissue of the buttocks and thighs. J Drugs Dermatol 2006;5:714–722.

27 Van Gemert MGC, Welch AJ: Time constant in thermal laser medicine. Lasers Surg Med 1989;9:405–421.

Michael Kreindel
Invasix Corp.
100 Leek Crescent, Unit 15
Richmond Hill, ON L4B 3E6 (Canada)
E-Mail mkreindel@invasix.com

Lapidoth M, Halachmi S (eds): Radiofrequency in Cosmetic Dermatology.
Aesthet Dermatol. Basel, Karger, 2015, vol 2, pp 23–32 (DOI: 10.1159/000362748)

Monopolar Radiofrequency

Katie Osley · Nicholas Ross · Nazanin Saedi

Department of Dermatology and Cutaneous Biology, Thomas Jefferson University, Philadelphia, Pa., USA

Abstract

Radiofrequency (RF) was first applied to aesthetic uses as monopolar RF. In its monopolar form, RF is applied to the target area by means of a dedicated handpiece, while a grounding pad is applied to the body at a distance. This differs from the mode of unipolar RF, in which a single electrode with no return is applied to the skin; the distinction between these two modalities is discussed in detail in the chapter 'Unipolar Radiofrequency'. The entry of RF into the body at the site of contact with the active electrode leads to bulk tissue heating, which can be provided at a depth. This ability to heat a volume of tissue noninvasively is applied to skin tightening on the face and body, as the heat induces both immediate collagen contraction and delayed collagen synthesis, by thermal induction of fibroblasts. Cumulative experience with monopolar RF has led to better, more reproducible, and better-tolerated results as treatment protocols and devices have evolved. Monopolar RF has a firm place as a safe and popular technology in the aesthetic armamentarium.

Nonablative radiofrequency (NARF) devices find their place in dermatology for the purpose of skin tightening as they readily target the deeper tissues of the dermis and induce collagen remodeling. They have gained particular popularity because of their ability to offer improvement of lax and/or photodamaged skin without the postoperative morbidity and financial burden of surgical procedures. Compared to more invasive alternatives, the results of NARF are modest; however, it remains in demand secondary to its lower side effect profile and remarkably short postprocedural downtime. This continuing shift away from ablative and invasive aesthetic procedures continues to be driven largely by patient and clinician preferences [1].

Monopolar radiofrequency (RF) is a nonsurgical approach to skin tightening and photorejuvenation. The FDA first cleared its use for treatment of periorbital rhytids in 2002. Additional clearances were given for tightening of the lower face in 2004 and for skin tightening on the body in 2005. The first device, Thermage ThermaCool (Sol-

Table 1. Monopolar radio frequency devices

Device	Manufacturer	Features
Thermage ThermaCool TC	Solta Medical	faster treatment tip
TC-3	Solta Medical	tip with faster discharge speed; larger tips
ThermaCool NXT (fig. 2)	Solta Medical	specific tips for body and eyes; handpiece for cellulite
Thermage CPT (Comfort Plus Technology)	Solta Medical	vibrating handpiece; thermal distribution control tip
Exilis	BTL Aesthetics, Prague, Czech Republic	dynamic Peltier cooling system; body and face handpiece
Pellevé S5 Wrinkle Treatment Generator (fig. 1)	Ellman International	GlideSafe handpiece with four sizes; progressive continuous heating

ta Medical Inc., Hayward, Calif., USA), was limited by patient discomfort, poorly reproducible results, and adverse events such as fat atrophy. However, improved treatment algorithms and modifications to the device have led to better patient satisfaction and results. Since the introduction of the approach, several devices utilizing monopolar RF have been marketed for various indications. A partial list of the commonly studied monopolar RF devices is presented in table 1.

Unlike light-based therapies based on selective photothermolysis, RF therapies do not provide selectivity based on a target chromophore. Rather, electrical current in the RF spectrum travels along tissue paths according to intrinsic resistance levels. Energy is delivered via a capacitative-coupled electrode to the skin. The device's alternating current shifts polarity of the electric field rapidly, forcing charged particles to flip their orientation. The tissue's natural resistance to this shift causes heat production, which in turn modifies micro- and macro-molecular structures [2–6]. This in turn induces 'volumetric heating', meaning that a three-dimensional volume of tissue is heated. Monopolar RF is applied with a single active electrode and a larger return electrode, applied at a distance from the treatment site. This differs from the mode of unipolar RF, in which a single electrode with no return is applied to the skin; the distinction between these two modalities is discussed in detail in the chapter 'Unipolar Radiofrequency'.

Patient Selection and Periprocedural Preparation

Proper patient selection for NARF is critical for optimal results. Clinicians must discuss expectations and realistic treatment outcomes prior to initiating the treatment sessions. This is particularly important for patients who have undergone prior surgical skin tightening, as their expectations typically reflect the more immediate and dramatic results of that modality.

Various attributes must be taken into account when selecting patients for this skin-tightening procedure. In general, those with mild to moderate skin laxity without underlying muscular or structural deformity will note greatest improvement. Specific body areas such as the eyelid, eyebrow, midface, melolabial folds, jowls, neck, thighs, arms, and décolletage tend to yield better results [7, 8]. The cumulative experience among providers and from the literature reveals that RF procedures appear to provide the best outcomes in persons aged 35–60 years, on crepe or wrinkled skin, and in individuals that are postpartum or after weight loss. Patients less likely to benefit from this procedure include those with severe solar damage or elastosis, obese patients, patients with fluctuating weights, and those in poor general or mental health. Moreover, patients on chronic corticosteroid or non-steroidal anti-inflammatory drugs may be poorly suited secondary to blunted healing responses [8].

The procedure has certain benefits over light-based approaches. Due in part to the epidermal sparing of NARF, this procedure can in theory be performed on all Fitzpatrick skin types, although darker skin may have more severe sequelae if burns occur. In addition, male patients can undergo this procedure with lower risk of loss of facial hair then laser or light-based treatments. Furthermore, the procedure can be performed on patients who have had prior rhytidectomies, blepharoplasties, or those who have had injectable toxins and fillers.

The main outcomes of treatment are both the physical changes and the patient's perception of the change. Therefore, it is recommended that the clinician document and demonstrate results for patients by taking standardized pre- and posttreatment photographs [9]. Photographs should be taken without makeup and jewelry and utilize standard backdrops, lighting, and distance. The use of a black headband to hold back long hair and a black cloak draped on the shoulders enhances standardization and draws attention to the relevant features. Frontal, left, right, and oblique (left and right) views should be taken at baseline (before treatment). These can be taken at 3 months' follow-up and at 6 months' follow-up to document response to treatment [10].

Certain periprocedural techniques can improve tolerability. Application of a topical anesthetic can alleviate the discomfort of superficial cooling provided by some devices. If needed, 'deeper' pain caused by the RF energy travelling through the subepidermal layers can be overcome by oral, intramuscular, or subcutaneous anxiolytics. Expert opinion recommends avoiding tumescent anesthetic as the water added to the target tissue can alter electrical conduction due to modification of physical properties of the tissue [11]. Another method of pain abatement is a verbal communication system between the patient and clinician to adjust treatment intensities. Typically, the provider will establish a 4-point scale, with 0 defined as no pain and 4 defined as maximum pain, to allow the patient to communicate pain levels throughout the procedure. By communicating current pain levels, the provider can make real-time adjustments of treatment application in order to minimize patient discomfort [7]. Pain management using all of these tools is best discussed with a prospective patient prior to embarking on treatment.

Technique

Once patients have been consulted regarding realistic expectations, informed consent should be obtained. Clinicians should advise patients that the treatment should feel hot but not unbearable; this clarifies what is expected during the treatment and establishes a definition for discomfort. Prior to positioning themselves on the treatment table, patients should remove all make-up and lotion to allow appropriate contact between the electrode and the skin. All jewelry must be removed and the patient's body should be entirely clear of any contact with metals or conductive surfaces in order to prevent undesired electrical transmission.

Monopolar RF, unlike unipolar or bipolar RF, requires the use of a grounding pad, which is attached to the patient's body with a temporary adhesive prior to the procedure. The monopolar device is configured to conduct the electrical current from the electrode on the handpiece to the grounding element at a distance. The main benefit of this approach is that it allows for deeper tissue penetration. The limitation is that this causes a high-energy density at the skin contact point; if not used correctly, the treatment can be painful [9]. Variables which impact the energy density, depth of penetration, and tissue effects are described in detail in the chapter 'Basic Radiofrequency: Physics and Safety and Application to Aesthetic Medicine'.

Some treatments are performed with prior skin marking to delineate the treatment area and map the placement of the handpieces. In the case of ThermaCool treatments, for example, the provider applies a ruled grid-work with transferrable ink to the skin, in which each square represents one 'spot'. With other devices, the handpiece may be moved continuously over the treatment area to achieve slow, progressive volumetric heating. Some providers introduce an infrared thermometer to monitor progress of the treatment in real time.

The monopolar RF procedure is performed with a coupling fluid or gel applied directly to the skin. This aids in achieving adequate and uniform contact between the electrode and the skin, which in turn improves uniformity of current transfer. In addition, the coupling medium provides hydration to the skin and thereby increases tissue conductivity. The handpiece is best applied to the skin with even pressure; uneven pressure can result in unequal energy application, which can harm the skin.

During treatment, the deep tissue heats while the skin surface remains cooler, particularly when a cooling element or gel is applied. If a cooling mechanism is not used, the handpiece must be used in motion or applied for short pulses to allow ambient cooling of the epidermis. Repeat pulses in the same area induce incremental heating of the dermis, while the epidermal cooling is required to prevent burns. As the dermis heats, and since the epidermis is allowed to cool, a thermal gradient develops. RF current preferentially travels to the path of least resistance, and lower resistance is offered by the warmer tissue below the epidermis. As a result, RF conduction to the dermis increases as the treatment progresses [8, 9].

Treatment approaches vary by device. Most involve a target total dose to be applied to a given surface area. The dose may be modified in sensitive body areas or over bony prominences. Protocols undergo testing and evolution in the hands of users, driven by cumulative experience. Protocols may include multiple passes, staggered passes, circular passes, and 'vector' passes that follow the desired line of tension. Treatments are generally completed when the tissue has been maintained at a target temperature for a given length of time or when the end points of erythema, edema and mild skin tightening are achieved [8, 12, 13].

Various technique modifications can improve patient comfort, safety, and treatment outcomes. For example, when the upper lip is treated, wet gauze can be placed between the teeth and lip to minimize discomfort [8]. For treatment of an area overlying a bony prominence, the clinician can shift the patient's tissue so that the maximum amount of subcutaneous fat distances the prominence from the treatment tip, thereby minimizing pain. If this is not possible, as when the forehead is treated, the power may be reduced. In areas of curvature, such as the jawline, the tissue can be elevated or pinched to provide a better, more even surface for contact with the tip [14].

Treating the eyelid requires a specialized smaller tip as well as placement of protective eye shields. Two to three drops of proparacaine or tetracaine can prepare the eye followed by lubricating ointment to improve comfort. It is generally recommended that the skin being treated be pulled away from the eye to allow the electrode to be placed outside the orbital rim.

Clinical Evidence

The majority of clinical studies utilize subjective patient satisfaction scores, as patients' perceived improvement typically outweighs the practitioners' scoring. However, a study by Iyer et al. [12] demonstrates histologic changes following NARF. The authors conclude that the desired skin tightening appears to be related to 'architectural improvement' which is manifested by a histometric increase in epidermal thickness and increased granular layer thickness at 3-month follow-up. This epidermal thickening is associated with more pronounced rete ridges. Dermal changes include reduction of solar elastosis in the papillary and upper reticular dermis and increase in normal-appearing elastic fibers at 3 months' follow-up [12]. Interestingly, the timing of perceived clinical benefit coincides with histopathological changes. Further, patients cited improved laxity specifically at 4–6 months coinciding with peak histologic collagen remodeling [15].

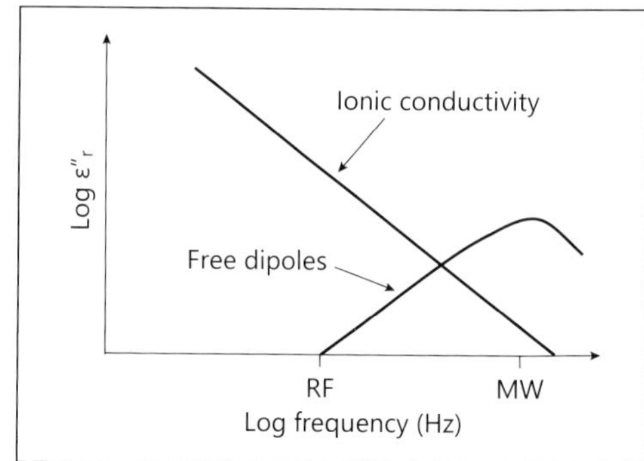

Fig. 1. Frequency dependence of dielectric loss factor (ε'').

The dielectric heating produces a heat dissipated in the unit of volume *(P)* that is proportional to the loss factor (the product of dielectric constant and loss tangent):

$$P = 2\pi f\, \varepsilon_o\, \varepsilon''r\; E^2\ W/m^3$$

where f = frequency (Hz), E = electric field strength (V/m), ε'' = dielectric loss factor, ε_o = permittivity of free space.

Practically speaking, this means that ionic currents in an alternating electromagnetic field diminish with increasing oscillating frequency. At a frequency where the amplitude of ion displacement is lower than λ_0, the ionic currents do not input energy into biological tissue. Water molecules can react on electromagnetic field oscillations at much higher frequencies; therefore, this heating dominates at high frequencies.

Penetration Depth

The shape of the thermal zone and penetration depth of electromagnetic energy depends on several factors:
- operating frequency
- geometry and configuration of the coupling system (electrode)
- input power
- treatment time
- treatment mode (stationary or motion).

Aesthetic procedures require that the skin not be overheated, and therefore are limited to operation with superficial penetration, less than 20 mm. The main task here is to concentrate a main volume of energy in the predetermined skin layers. For example, skin tightening requires energy deposition mostly in dermal area, whereas body contouring and fat destruction require deeper penetration of electromagnetic energy. Medical diathermy and physiotherapy procedures require deeper energy for reaching muscles, etc.

At lower frequencies, where heating is stipulated by displacement currents, an increase in frequency leads to decreasing penetration depth because of skin effect [2]. At low frequencies (below several MHz), it is possible to control operating frequency and consequently to change the penetration depth. For example, one marketed device uses three frequencies (0.8, 1.7, and 2.45 MHz) in order regulate a heating zone. The deepest penetration corresponds to lower frequency. In this case, it is necessary to correlate an operating frequency with radiofrequency (RF) power because of high divergence in the electromagnetic field at lower frequency. The penetration depth Dp as a result of dielectric heating can be calculated by the equation [3]:

$$\mathrm{Dp} = \frac{c\sqrt{\varepsilon'}}{2\pi f \varepsilon''}, \tag{1}$$

where ε' = relative dielectric permittivity and c = velocity of electromagnetic wave.

It can be derived from the equation below that the penetration depth of dielectric heating decreases with increased frequency, as in the case of conductive heating.

The temperature rise θz at depth z is given by:

$$\theta z = \theta_0 \exp(-z/\mathrm{Dp}) \tag{2}$$

It should be emphasized that penetration depth of energy as a result of dielectric heating is not the same as a skin depth [4]. The penetration depth increases essentially with the dielectric loss factor ε' that leads to power dissipation in accordance with equation 1. Because of this power dissipation, the power flux falls as the wave propagates into biological tissue.

Practically, the penetration depth depends on the electrode system configuration and the electrode shape. Usually, electrodes do not create a homogeneous electric field in biological tissue. The field strength is highest near coupling electrodes and strongly decreases between electrodes. Therefore, the maximum heating develops in proximity to the electrodes.

In the practice of aesthetic medicine, most systems operate at low frequencies with conductive current heating mechanism (Syneron, Endymed, and Venus devices: 1 MHz; Lumenis: 0.5 MHz; Viora 0.8, 1.7 and 2.45 MHz; Thermage: 6.78 MHz). For therapeutic and physiotherapeutic treatments, dielectric heating is most appropriate. Usually, systems operate at 27.13 MHz. The dynamics of these treatments is different than aesthetic procedures. The treatments are stationary, and penetration depth is greater, but RF power densities are lower. Accent systems produced by Alma Lasers operate at 40.68 MHz. The dielectric heating is dominant at this frequency, providing a possibility of deep homogeneous heating up to 20- to 25-mm depth. It should be mentioned that at high frequencies (more than several MHz), it is necessary to operate at frequencies permitted for industrial, scientific, and medical applications (ISM frequencies).

Methods of Radiofrequency Energy Coupling

There are several coupling methods for applying RF power to biological tissue.

(1) In most aesthetic devices, resistive biological tissue is brought in contact with a metal electrode. The disadvantages of this method are possible sparking due to poor contact as well as nonhomogeneous distribution of current density at frequencies higher than 5–10 MHz because of the skin effect on the electrodes.

(2) Capacitive coupling is more applicable at high frequencies. The electrodes are covered by a dielectric coating that produces a capacitor at the contact with biological tissue. The thickness of this barrier depends on dielectric purposes of the coating and operating frequency. At lower frequencies, the higher capacitance of this barrier is needed for efficient coupling.

(3) The inductive coupling methods operate by noncontact induction of electromagnetic field in the biological tissue. RF current flow through a coil creates an electromagnetic field. This can efficiently heat a biological tissue in the field. The advantage of this method is noncontact heating; the disadvantages are low heating intensity and low uniformity. Usually, these methods are more applicable for long-term procedures like heating for a physiotherapeutic effect.

(4) Radiative or antenna heating may be implemented by radiation of RF energy into biological tissue through an antenna dipole. This method is applicable at high frequencies where antenna dimensions begin to be practical.

(5) In resonant coupling, the electrode system forms a high-Q-factor resonator. The coupled biological tissue may be described by losses of this resonator. For example, the unipolar handpiece electrode system by Alma Lasers is a resonator with a Q-factor greater than 1,000 [5]. The equivalent resistance of this resonator is 15–20 kΩ. The coupled biological tissue can be described by a resistance of 250–350 Ω at 40.68 MHz. At this frequency, tissues demonstrate low capacitance (several pF). In this system, RF energy delivered to this resonator will be practically dissipated in the coupled biological tissue.

The electrode structure for resistive, capacitive or resonant coupling may be bipolar, monopolar, or unipolar. At low frequencies, most systems have a bipolar structure. All systems that claim that they are multipolar or multielectrode (for example TriPolar, Pollogen) are actually bipolar systems with re-switching of electrode pairs.

At higher frequencies, bipolar structures can be used as well. For example, the coaxipolar (bipolar) electrode structure of Accent (Alma Lasers) exploits two coaxial electrodes (outer grounded and central-energized electrode). In this case, the penetration depth depends on the distance between electrodes: greater interelectrode distance leads to deeper penetration (fig. 2).

It can be seen from this image that the penetration depth of the bipolar (coaxipolar) system is limited by the influence of a ground electrode and does not exceed 7–8 mm. The heating is maximal at the central area between electrodes and practically absent below the electrodes.

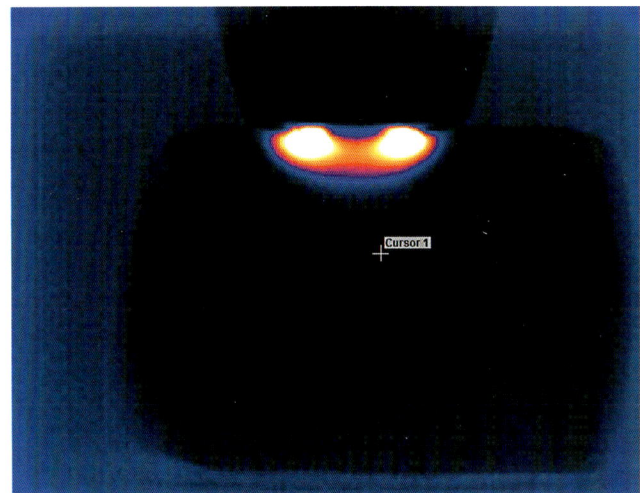

Fig. 2. Thermal image of bipolar heating of biological tissue.

At higher frequencies (above 6 MHz, such as Thermage), it is possible to operate with monopolar electrodes. Monopolar electrode systems actually use two electrodes: an active or energized electrode and a ground pad. The current density in the area of ground electrode is very low; therefore, the heating of biological tissue comes from the energized electrode. The penetration depth is enough for heating dermal area. Nevertheless, at this frequency the dielectric heating is weak, and penetration depth that is defined mostly by skin effect should be very limited.

Unipolar electrode systems can operate efficiently at frequencies higher than 13.56 MHz. The prevailing mechanism of heating at 40.68 MHz is dielectric heating. The penetration depth is 15–20 mm. The advantage of this system is the absence of a second electrode. The distortion of the heating zone related to the second electrode's influence is minimal. The maximal heating is at the central axis of the electrode system, as seen in figure 3.

The unipolar electrode system of the Accent system (Alma Laser) uses spherical electrodes. The divergence of electromagnetic field limits the penetration depth at values applicable for an aesthetic procedure such as body contouring and skin tightening.

Radiofrequency Power and Its Regulation

Penetration depth and temperature change dynamics depend on applied RF power and treatment time. Different systems have different levels of operating RF power from 20 up to 300–400 W for the systems at high frequencies. The input RF power also depends on the coupling electrodes. For example, the Alma Accent has operating

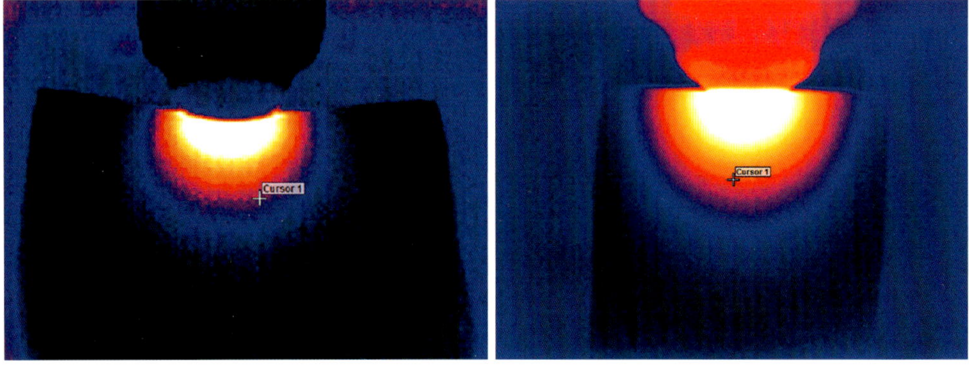

Fig. 3. Thermal images of unipolar heating at different RF power levels.

average RF power of up to 300 W because of the high penetration depth of RF energy. The Thermage monopolar system has a similar power level.

RF power level also depends on the treatment method. If the electrodes are placed and held in place, i.e. a stationary treatment is performed, the energy should be limited by RF power or treatment time. The technique of in-motion treatment provides application of high power over large skin areas during a continuous treatment. The disadvantage of this type of treatment is operator-dependent variability.

Output power control can be implemented by amplitude control or pulse width modulation (PWM) control. The PWM control is modulation with rectangular pulses. The amplitude of output power and frequency of modulation are kept constant, but the pulse duration or duty cycle of pulses can be varied. As a result, average power will be controlled while allowing high peak power. Since the cooling of biological tissue is slow, if the frequency of RF modulation is high enough, the tissue heating will not depend on this frequency. Rather, the biological tissue will reach a thermodynamic equilibrium (in which the heating rate is equal to the cooling rate) at a certain temperature, depending on average RF power. The frequency of the modulation varies with different devices. In practice, a frequency of 15 kHz is used in order to avoid an audio effect of modulation.

Impedance Matching

Biological tissue can be simulated as a resistor and capacitor in parallel. The impedance of this circuit and, consequently, biological tissue, is different from 50 Ω. If an RF generator or amplifier that is usually matched with 50 Ω operates with such a load (biological tissue), there will be high reflection of RF power back to the generator. The effective power that is dissipated as heat in the biological tissue will be low. Moreover, RF generators (amplifiers) have to withstand high levels of reflection power; this need increases the requirements of reliability in RF power sources.

Impedance matching systems convert an intrinsic impedance of biological tissue to 50 Ω. Usually, impedance matching systems at RF frequencies are a combination of several capacitors and inductors. The challenge in these devices is that they must distinguish the different impedances of different parts of the human body. The impedance matching system must compensate for these differences. An impedance matching system may be variable (Thermage) or broadband (Accent). A variable impedance matching system is suitable for stationary treatments because the process of impedance correction takes time. For systems which are operated in motion, like Accent, broadband impedance matching is more applicable.

Phase Control of Penetration Depth

Phase control is used in the unipolar devices of the Alma Accent system. This is possible because the reflected power measured between the RF power amplifier (electrode) and the impedance matching system is less than 1–3% of the incident RF power. A value termed the SWR coefficient, which characterizes a standing wave, is close to 1. This means that the travelling electromagnetic wave propagates between the RF amplifier and the impedance matching system. Furthermore, it means that RF voltage is equivalent at all points along the RF path.

The voltage U alters in time in accordance with:

$$U = A\sin(2\pi ft - \phi)$$

where A = amplitude of voltage and ϕ = initial phase.

This equation means that the voltage reaches the same amplitude periodically at all points along the RF path at different times. A standing wave is sustained between the impedance matching system and the electrode; therefore, at any given point the voltage oscillates between 0 (the node) and twice the amplitude (antinode).

If a phase shifter is placed between the impedance matching system and the electrode, i.e. in the travelling wave portion of the RF path, it will change the electrical length of the RF path and consequently the phase of the electromagnetic wave. This makes it possible to regulate RF voltage at the contact area of the electrode and the skin. At low voltage, the penetration depth will be maximal, whereas at high voltage the RF energy will be greatest in the skin (fig. 4).

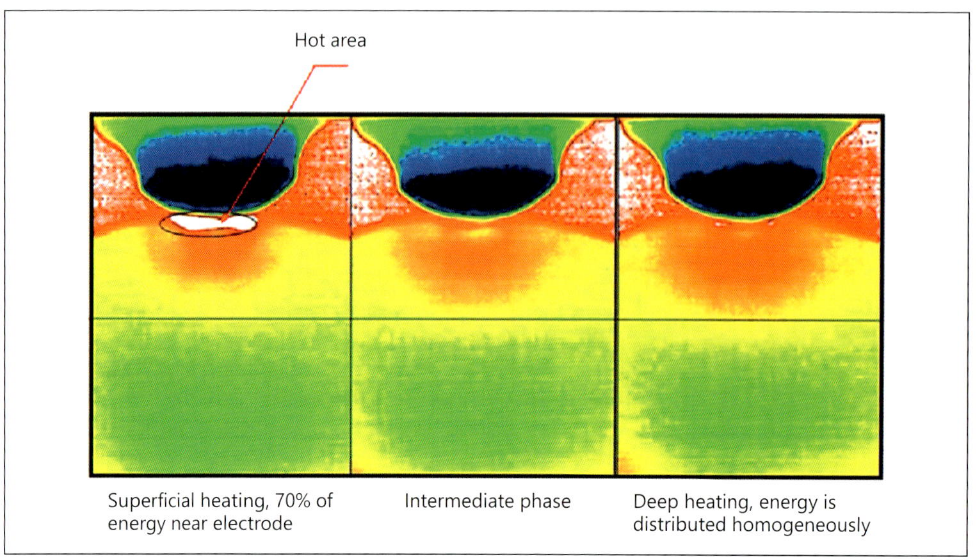

Fig. 4. Thermal images of unipolar heating at different phases of electromagnetic wave.

Cooling and Inverse Thermal Gradient

Practically, all RF systems operating with high RF energy volumetric density and high penetration depth (more than several mm) have integral cooling of treated tissue. It was discovered that a reverse thermal gradient provides controlled contraction of collagen tissue by stimulation of intensive neocollagenesis [6].

Thermage uses cryogenic cooling of the treated tissue before and after treatments. This intensive cooling provides a necessary thermal gradient and decreases pain effects. This approach is suitable for stationary systems. When the treatment is done in motion, continuous cooling is needed. Motion devices such as Accent include cooling within the electrode.

Unipolar RF: Clinical Effects

Unipolar RF can be used for skin tightening (mainly face and neck) or body shaping (cellulite and circumference reduction). The mechanism of action is attributed to the high-frequency electromagnetic radiation that induced rotational oscillations in the dermal water molecules that eventually will create heat and collagen regeneration. The heat can get up to a depth of 20 mm.

Goldberg et al. [7] evaluated the efficacy of a unipolar device (Accent RF System; Alma Lasers, Buffalo Grove, Ill., USA) in 30 patients with grade III or IV cellulite on the upper thigh. Patients underwent 6 treatments at 2-week intervals. Twenty-seven of 30 patients treated experienced clinical improvement 6 months after treatment, with a mean decrease of 2.45 cm in thigh circumference. Minimal side effects were reported. Histological evidence of dermal fibrosis was reported, but there was no magnetic resonance imaging evidence of changes in the pannicular layer. The authors propose that RF-induced contraction between the dermis and Camper's fascia may explain the initial skin tightening effect but that its longer-lasting effect is due to dermal fibrosis. Del Pino et al. [8] reported 20% contraction between the stratum corneum and Camper's fascia in 68% of patients 15 days after treatment using a unipolar RF device (Accent RF System). Goldberg and colleagues did not find this effect 6 months after treatment, indicating that it may be a transient response. Alexiades-Armenakas et al. [9] did a randomized, blinded, split-design, controlled study on 10 individuals with grade II–IV cellulite and similarly reported favorable results of clinically visible and quantifiable improvement of cellulite 3 months after treatment. Quantified improvement did not reach statistical significance but showed a trend toward improvement that was observed in all patients following a mean of 4 treatments at 2-week intervals using a unipolar RF device [9]. In another study, same authors did randomized, split-face study with blinded evaluations that demonstrated that minimal pass, mobile energy delivery serial treatments with either the unipolar or bipolar handpieces of a novel RF device appear to be safe and painless. Each handpiece demonstrated minimal clinical efficacy which was not statistically significant, but again, with a trend toward improvement in rhytides and laxity of facial skin [10].

Conclusion

Although often described together, unipolar RF differs from monopolar RF in several aspects. Unipolar RF systems can be effectively used for heating tissue due to these distinctions. The heating zone is determined by the operating frequency, dissipated energy, and electrode configuration. At frequencies higher than 27.13 MHz, the dominant mechanism of action by RF is dielectric heating. Operation at high frequencies is more appropriate for body contouring, because higher frequencies can provide greater penetration depth. Operating at high frequency decreases RF current density near the electrodes and allows operation at higher power levels without burning the skin. The method of RF coupling further impacts the RF-tissue interaction. Capacitive and resonant coupling of RF energy provides high efficiency of energy deposition.

In order to increase input energy, the RF-coupled system should be well matched with the treated biological tissue. Proper matching decreases reflected RF, and the incident RF will be more efficiently absorbed by biological tissue. Additional means to improve RF penetration depth include phase shifting. This novel approach changes the electrical length of RF current paths and allows control of RF voltage at the surface and depth of a biological tissue. By optimizing all of these parameters, RF devices can be designed for a wide range of tissue effects.

References

1 Hasted JB: Aqueous Dielectrics. London, Chapman and Hall, 1973.
2 Vorst AV, Rosen A, Kotsuka Y: RF/Microwave Interaction with Biological Tissues. New York, John Wiley & Sons, 2006.
3 von Hippel AR: Dielectric Properties and Waves. New York, John Wiley, 1954.
4 Meredith RJ: Engineer's Handbook of Industrial Microwave Heating. London, Institution of Electrical Engineers, 1998.
5 Karni Z, Britva A: System and method of heating biological tissue via RF energy. US patent 7630774. Dec 2009.
6 Knowlton EW: Method and apparatus for controlled contraction of collagen tissue. US patent 5,660,836. 1997.
7 Goldberg DJ, Gazeli A, Berlin AL: Clinical, laboratory, and MRI analysis of cellulite treatment with a unipolar radiofrequency device. Dermatol Surg 2008;34:204–209.
8 Del Pino E, Rosado RH, Azuela A, Graciela Guzmán M, et al: Effect of controlled volumetric tissue heating with radiofrequency on cellulite and the subcutaneous tissue of the buttocks and thighs. J Drugs Dermatol 2006;5:714–722.
9 Alexiades-Armenakas M, Dover JS, Arndt K: Unipolar radiofrequency to improve the appearance of cellulite. Dermatol Surg 2008;10:148–153.
10 Alexiades-Armenakas M, Dover JS, Arndt KA: Unipolar versus bipolar radiofrequency treatment of rhytides and laxity using a mobile painless delivery method. Lasers Surg Med 2008;40:446–453.

Moshe Lapidoth, MD, MPH
Dermatology Laser Unit, Dermatology Department
Rabin Medical Center
7 Keren Kayemet St.
Petach Tikva (Israel)
E-Mail alapidot@netvision.net.il

Halachmi · Britva · Lapidoth

Lapidoth M, Halachmi S (eds): Radiofrequency in Cosmetic Dermatology.
Aesthet Dermatol. Basel, Karger, 2015, vol 2, pp 43–49 (DOI: 10.1159/000362766)

Bipolar Radiofrequency

Shimon Eckhouse[a] · Meltem Onder[b, c] · Klaus Fritz[b, d]

[a] Syneron-Candela, Yokneam, Israel; [b] Dermatology and Laser Center, Landau, Germany; [c] Department of
Dermatology, Faculty of Medicine, Gazi University, Ankara, Turkey; [d] Carol Davila University, Bucharest, Romania

Abstract

Bipolar radiofrequency (RF) utilizes symmetric and closely placed electrodes to pass RF current
through a defined volume of tissue. The approach has benefits over monopolar and unipolar RF in
its control. To optimize these, the electrodes and the RF parameters must be designed and operated
in a manner that heats the desired target. The applications of bipolar RF are wide and range from
medical applications such as treatment of acne to aesthetic applications in the skin and subcutaneous tissue. This chapter reviews the clinical literature on application of bipolar RF on various aesthetic indications.

© 2015 S. Karger AG, Basel

All radiofrequency (RF) treatments rely on the passage of current between a positive
and negative electrode. In 'monopolar' RF, a single electrode is placed at the target site
and a distant return electrode is placed on a distant body site. In contrast, bipolar RF
applies two electrodes in close proximity at a fixed distance from each other. By plac-
ing the electrodes close together, the current is limited to the tissue between the electrodes, allowing greater control over the size of the treated volume, but usually at the
cost of depth of penetration which monopolar RF allows. Further, in contrast to monopolar RF, the electrodes of bipolar RF are generally symmetric, which allows sym-
metrical treatment. Additional advantages of bipolar RF are reduced energy loss, due
to the proximity of the electrodes, and reduced energy density at the treatment area,
with subsequent reduction in the risk of overheating and burning of the skin below
the return electrode [1]. The features of bipolar RF have also allowed the development
of fractional bipolar RF technology, in which a matrix of electrodes creates microfoci
of epidermal ablation with underlying dermal heating in a manner conceptually sim-

ilar to fractional lasers. This approach is discussed in the chapter by Eckhouse et al. [pp. 70–80].

Due to the greater control over energy density and pattern of deposition, bipolar RF devices are better tolerated and cause less pain. They are also more suitable for homeostasis and controlled vessel contraction [1–3]. Similar to the mechanism of action of monopolar RF devices, the electric current generates heat and meets resistance from the tissue. This heat causes collagen shrinkage, an inflammatory response, and contraction in the dermis. The limitation of current to the area between the electrodes also makes these systems safer in patients with implanted devices.

Bipolar RF devices can be combined with simultaneous light therapies for a synergistic effect. This approach, which underlies the technologies branded as electro-optical synergy (elōs; Syneron Medical Ltd., Yokneam, Israel), provides two heating modalities in parallel. This approach allows dose reduction of each modality. This approach can reduce risk of adverse events, as each modality provides a lower amount of heating. More important, the selective heating of a target by the light (laser or intense pulsed light) component of elōs leads to preferential heating of the target by RF as electric currents travel in the path of least resistance – in this case, in the warmer tissue targeted by the light selectively. This application of synergistic energy sources is discussed further in the chapter by Eckhouse [pp. 70–80].

Studies and Evidence-Based Effectiveness

Skin Rejuvenation

The most studied bipolar RF devices for this indication use elōs (electro optical synergy) with broadband light (Aurora, Syneron Medical) or with diode laser (Polaris, Syneron Medical) [2]. Both devices utilize bipolar RF to elicit deeper tissue heating in order to induce neocollagenesis.

Yu et al. [4] evaluated the safety and efficacy of elōs in skin tightening in 19 Chinese patients with skin laxity and periorbital rhytides. This is the first study to demonstrate the effect of broadband infrared (IR) light plus bipolar RF in the treatment of facial wrinkles in Asians. After 3 months, a significant improvement was observed in skin laxity of the periorbital, cheek, jowl, and upper neck areas. There were no serious complications. El Domyati et al. [5] evaluated histological changes following treatment. Six volunteers were treated over the periorbital region, and punch biopsies were taken after six treatment sessions. Histologic analysis revealed an increase in epidermal thickness, a 53% reduction in elastin content and 28% increase in collagen fibers. Sadick et al. [6] reported similar results. Overall, average wrinkle improvement was 41.2%, while pore size and pigmentation improved by 65%. This approach was successfully used to treat deep wrinkles and photoaging in several studies without sig-

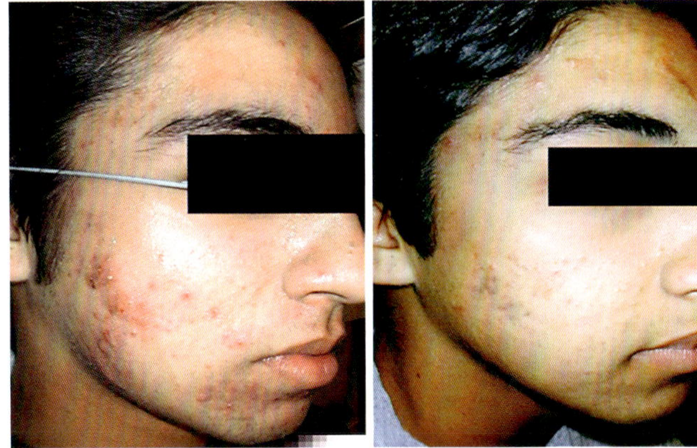

Fig. 1. Treatment of acne; left: before, right: after. Courtesy Dr. Tess Mauricio, USA.

nificant side effects [7, 8]. Hammes et al. [9] reported a notable improvement in skin rejuvenation. They suggested that the clinical effect may be caused by matrix coagulation of the deep dermis by the RF and selective thermolysis of blood vessels with the 900-nm diode laser.

FACES (Functional Aspiration Controlled Electrothermal Stimulation) is another technology used with a bipolar RF device that employs a vacuum to maximize and control penetration (ALUMA System Lumenis Inc., Santa Clara, Calif., USA) [10]. One study using this system included 30 patients with periocular wrinkles, glabellar wrinkles, laxity of the cheeks with accentuation of the nasogenian furrow, striae distensae at the scapulohumeral joint, abdomen and gluteal-trochanteric areas, or acne scars. Study participants underwent a cycle of 6–8 sessions in 2-week intervals with the bipolar RF device. All patients showed improvement in treated imperfections from the second session onward, and they expressed their satisfaction at the end of the treatment cycle. The most notable clinical, histological and immunohistochemical results were observed in the patients with abdominal striae distensae. Gold et al. [11] used this system in 46 patients as well and found significant improvements in facial wrinkles.

Acne Vulgaris

Bipolar RF combined with optical energies may be an alternative nonablative modality for the treatment of moderate acne vulgaris. Prieto et al. [12] used the Aurora (Syneron Medical) device in 32 patients with moderate acne, achieving a 47% reduction of lesions, a decrease in the size of sebaceous glands as well as a reduction in perifollicular lymphocytic infiltrates. Clinical improvement can be significant (fig. 1). Treatment of acne may require a series of treatments followed by maintenance treatments.

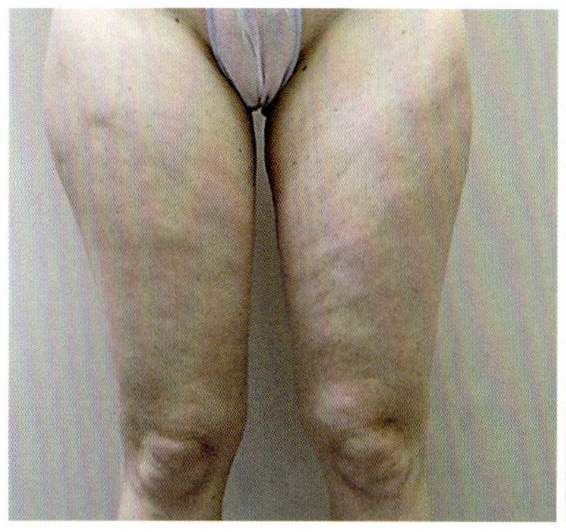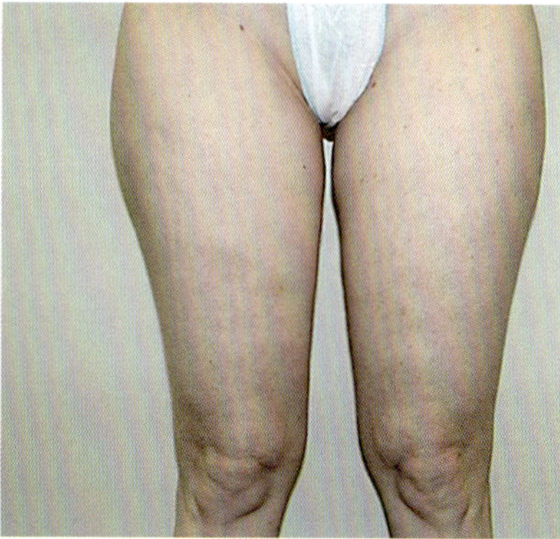

Fig. 2. Improvement in the appearance of cellulite; left: before, right: after. Courtesy Dr. Gerald Boey, Canada.

Cellulite

The deep heating of tissue using bipolar electrodes has previously been achieved with the combination of RF and IR light. This approach is also used for the treatment of cellulite. One approach to cellulite combines bipolar RF with IR light (700–2,000 nm), vacuum, and massage (VelaSmooth and VelaShape systems, Syneron Medical) for the treatment of cellulite. The IR light and RF heat the skin and fat layer respectively, and while heat and massage stimulate lymphatic flow and strengthen the fibrous network. The heat also causes thermal damage to surrounding adipose tissue penetrating 3 mm deep [13]. In a pilot study, 12 women with cellulite were treated twice weekly for a total of 8–9 treatments [14]. Results showed a beneficial effect on the reduction of abdomen and thigh circumference, and a smoothening of cellulite. In Romero's study, improvements in the overall cellulite appearance and skin condition were achieved. The pain during treatment was reported as being bearable by all patients (no patients scored more than 5 points on an 11-point visual analogue scale) and the erythema lasted a few hours [15]. Sadick and Mulholland [16] treated 35 cellulite patients twice weekly using the same device. After 16 treatments, skin smoothing, cellulite improvement and reductions in thigh circumference were seen. Alster and Tanzi [17] performed a controlled 20-patient study and reported a 50% improvement in cellulite with 90% patient satisfaction. The results can be seen after a series of 5–8 treatments (fig. 2). The FDA-cleared indication for cellulite is 'temporary improvement in the appearance of cellulite'. In keeping with this, all cleared devices do, in practice, require maintenance treatments to retain the effect.

ThermaLipo (Thermamedic Ltd., Alicante, Spain) is a bipolar RF device that uses a sensor that constantly monitors the return current as an indicator of impedance and adjusts the frequency to achieve total tissue volume penetration, while maintaining deeper tissue temperatures to ensure electrothermal damage deposition. The pulsed-

Eckhouse · Onder · Fritz

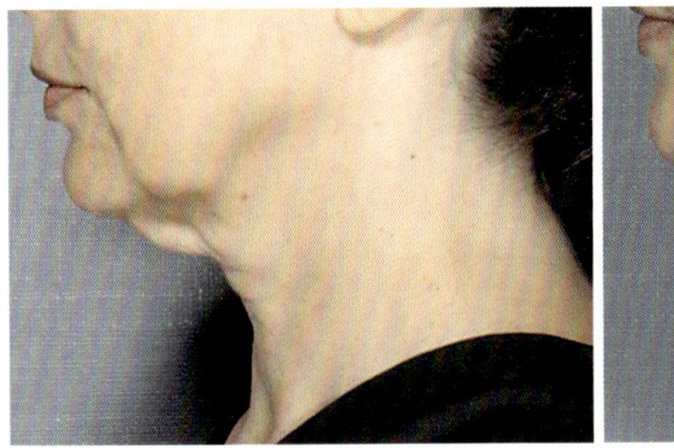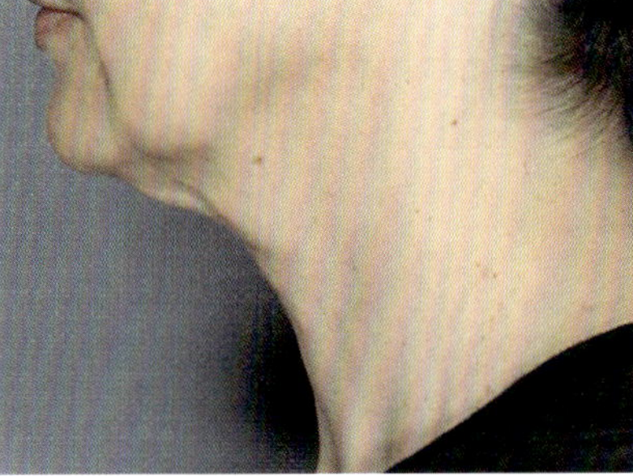

Fig. 3. Improvement in neck and jawline laxity; left: before, right: after 5 treatments. Courtesy Dr. Roy Geronemus, USA.

mode RF has the ability to deposit a high energy load rapidly, which increasingly raises the temperature of the subcutaneous tissue and skin. In one study conducted by van der Lugt et al. [18], 50 patients with cellulite were treated. Results showed a clear overall improvement of the shape of the buttocks, and less evidence of the usual cellulite-associated dimpled appearance was seen.

Upper Arm, Neck and Jaw Line

A study performed by Brightman et al. [19] demonstrated the benefits of the combination of IR, bipolar RF, vacuum, and mechanical massage. Its efficacy and safety when evaluated on upper arms, abdominal and flank circumferences showed significant reduction in circumference and improvement in the appearance of the arms and abdomen.

Belenky et al. [20] evaluated recent developments and trends in RF technology. A bipolar RF device (Reaction Viora Inc., Jersey City, N.J., USA) combined with a mechanical massage technique was used. They focused on the effect of RF on the neck and jaw line and found that a moderate improvement could be achieved in the treatment of jowls.

Conclusions

The popularity of noninvasive techniques and technologies that can address the cutaneous signs of skin aging and other common cosmetic indications is on the rise. Novel bipolar RF energy-based devices represent one such technology that has been shown to achieve improvements in these cosmetic indications safely and effectively. Nonablative skin rejuvenation with bipolar RF achieves skin tightening through controlled dermal collagen contraction, neocollagenesis, and neoelastogenesis. One of the central advantages of bipolar RF-based systems is that treatment results in minimal morbidity and a low risk for postprocedural complications with little to no downtime, particularly when compared with more invasive cosmetic techniques, offering patients a noninvasive, nonsurgical treatment alternative.

The combined simultaneous use of RF energy, IR energy, mechanical massage, and suction is an innovative and promising treatment approach that appears to have several potential and growing applications in aesthetic medicine. The combination of these modalities has also been assessed in patients of darker Fitzpatrick skin types, and found to carry a lower risk of adverse events such as postinflammatory hyperpigmentation. Cosmetic treatments with bipolar RF devices are widely characterized as being well tolerated, and appear to be better tolerated than monopolar RF systems. Though the literature has shown that bipolar RF devices are effective in improving numerous cosmetic indications including skin rejuvenation, skin tightening, acne vulgaris, and cellulite, the best results have been reported with the efficacy of the elōs system for the treatment of cellulite.

References

1 Lolis MS, Goldberg DJ: Radiofrequency in cosmetic dermatology: a review. Dermatol Surg 2012;38: 1765–1776.
2 Atiyeh BS, Dibo SA: Nonsurgical nonablative treatment of aging skin: radiofrequency technologies between aggressive marketing and evidence-based efficacy. Aesthetic Plast Surg 2009;33:283–294.
3 Alster TS, Lupton JR: Nonablative cutaneous remodeling using radiofrequency devices. Clin Dermatol 2007;25:487–491.
4 Yu CS, Yeung CK, Shek SY, Tse RK, Kono T, Chan HH: Combined infrared light and bipolar radiofrequency for skin tightening in Asians. Lasers Surg Med 2007;39:471–475.
5 El-Domyati M, El-Ammawi TS, Medhat W, Moawad O, Mahoney MG, Uitto J: Electro-optical synergy technique: a new and effective nonablative approach to skin aging. J Clin Aesthet Dermatol 2010;3:22–30.
6 Sadick NS, Alexiades-Armenakas M, Bitter P Jr, Hruza G, Mulholland RS: Enhanced full-face skin rejuvenation using synchronous intense pulsed optical and conducted bipolar radiofrequency energy (elōs): introducing selective radiophotothermolysis. J Drugs Dermatol 2005;4:181–186.
7 Kulick M: Evaluation of a combined laser-radio frequency device (Polaris WR) for the nonablative treatment of facial wrinkles. J Cosmet Laser Ther 2005;7:87–92.
8 Doshi SN, Alster SN: Combination radiofrequency and diode laser for treatment of facial rhytides and skin laxity. J Cosmet Laser Ther 2005;7:11–15.
9 Hammes S, Greve B, Raulin C: Electro-optical synergy (elōs) technology for nonablative skin rejuvenation: a preliminary prospective study. J Eur Acad Dermatol Venereol 2006;20:1070–1075.
10 Montesi G, Calvieri S, Balzani A, Gold MH: Bipolar radiofrequency in the treatment of dermatologic imperfections: clinicopathological and immunohistochemical aspects. J Drugs Dermatol 2007;6:890–896.
11 Gold MH, Goldman MP, Rao J, Carcamo AS, Ehrlich M: Treatment of wrinkles and elastosis using vacuum-assisted bipolar radiofrequency heating of the dermis. Dermatol Surg 2007;33:300–309.
12 Prieto VG, Zhang PS, Sadick NS: Evaluation of pulsed light and radiofrequency combined for the treatment of acne vulgaris with histologic analysis of facial skin biopsies. J Cosmet Laser Ther 2005;7: 63–68.
13 Wanitphakdeedecha R, Manuskiatti W: Treatment of cellulite with a bipolar radiofrequency, infrared heat, and pulsatile suction device: a pilot study. Cosmet Dermatol 2006;5:284–288.
14 Hexsel DM, Siega C, Schilling-Souza J, Porto MD, Rodrigues TC: A bipolar radiofrequency, infrared, vacuum and mechanical massage device for treatment of cellulite: a pilot study. J Cosmet Laser Ther 2011;13:297–302.
15 Romero C, Caballero N, Herrero M, Ruíz R, Sadick NS, Trelles MA: Effects of cellulite treatment with RF, IR light, mechanical massage and suction treating one buttock with the contralateral as a control. J Cosmet Laser Ther 2008;10:193–201.

16 Sadick NS, Mulholland RS: A prospective clinical study to evaluate the efficacy and safety of cellulite treatment using the combination of optical and RF energies for subcutaneous tissue heating. J Cosmet Laser Ther 2004;6:187–190.

17 Alster TS, Tanzi EL: Cellulite treatment using a novel combination radiofrequency, infrared light, and mechanical tissue manipulation device. J Cosmet Laser Ther 2005;7:81–85.

18 van der Lugt C, Romero C, Ancona D, Al-Zarouni M, Perera J, Trelles MA: A multicenter study of cellulite treatment with a variable emission radio frequency system. Dermatol Ther 2009;22:74–84.

19 Brightman L, Weiss E, Chapas AM, Karen J, Hale E, Bernstein L, Geronemus RG: Improvement in arm and post-partum abdominal and flank subcutaneous fat deposits and skin laxity using a bipolar radiofrequency, infrared, vacuum and mechanical massage device. Lasers Surg Med 2009;41:791–798.

20 Belenky I, Margulis A, Elman M, Bar-Yosef U, Paun SD: Exploring channeling optimized radiofrequency energy: a review of radiofrequency history and applications in esthetic fields. Adv Ther 2012;29:249–266.

Meltem Onder
Laser and Dermatology Center
Ankara, 76829 (Turkey)
E-Mail meltemonder@yahoo.com

Lapidoth M, Halachmi S (eds): Radiofrequency in Cosmetic Dermatology.
Aesthet Dermatol. Basel, Karger, 2015, vol 2, pp 50–61 (DOI: 10.1159/000362767)

Fractional Radiofrequency

Shlomit Halachmi[a] · Meltem Onder[b, c] · Klaus Fritz[b, d]

[a]Herzelia Dermatology and Laser Center, Herzelia Pituach, Israel; [b]Dermatology and Laser Center, Landau,
Germany; [c]Department of Dermatology, Faculty of Medicine, Gazi University, Ankara, Turkey; [d]Carol Davila
University, Bucharest, Romania

Abstract
Fractional radiofrequency (FRF) was developed with the same conceptual framework as fractional
lasers, which is to provide focal, high-energy treatment zones within intact skin for the purpose of
reduced downtime and risk. Since the heating profile of radiofrequency can in some cases be controlled with greater precision than laser, this technology has gained popularity and is perceived as
a safer alternative for some applications. FRF can be applied with a variety of applicator geometries,
which in turn affect the skin response. This chapter highlights the variants of FRF and summarizes
the literature for aesthetic applications.   © 2015 S. Karger AG, Basel

Fractional bipolar radiofrequency (RF) devices are gaining popularity in aesthetic
medicine. Initial devices were monopolar or bipolar, technologies reviewed in the
chapters 'Monopolar Radiofrequency' and 'Bipolar Radiofrequency'. Both techniques are used for delivering RF energy to the skin for the purpose of heating the
dermis and, in some cases, the epidermis. Both technologies were developed for
application in nonablative modalities, primarily in skin tightening and body contouring.

As these devices were being developed, another branch of energy-based devices
was being modified for a new use: fractional delivery of laser light [1]. Fractional lasers
were developed to address the drawbacks of 'confluent' resurfacing, which was limited by significant downtime and high risks of infection, pigmentary change, and scarring. While full-face resurfacing was very effective, the risk-benefit profile was too
great for many patients and providers. Fractional lasers were designed to improve the

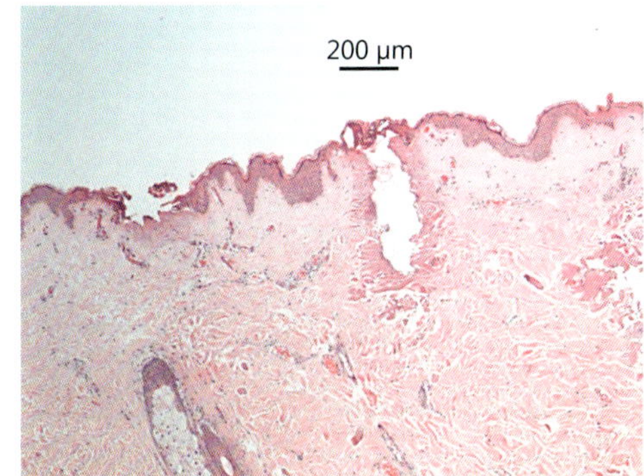

Fig. 1. Histology of fractional laser defect in skin, showing ablation surrounded by coagulation. The defect is widest at the surface and tapers with depth, forming a 'V'-shaped wound.

risk-benefit profile. By delivering concentrated energy in very small foci, the treatment delivered energy to deep epidermis and dermis while sparing much of the epidermis. The foci of epidermal damage were surrounded by untreated epidermis, which allowed very rapid reepithelialization, on the order of 3 days, with resulting reduction in downtime, infection rates, scarring, and pigmentary changes. The initial nonablative lasers were soon complemented with ablative lasers, which could provide results more similar to classic resurfacing.

Fractional lasers remain extremely popular in both the ablative and nonablative modalities. However, all laser-based devices are subject to the laws of light-tissue interaction, and as a result of unavoidable absorption in the topmost tissue layers, the amount of light that penetrates deeper into the dermis is progressively reduced. This leads to the classic 'V'-shaped defect observed in histology (fig. 1). RF has the potential to provide different patterns of energy and heat distribution. The desire for fractional therapies with increased options for tissue heating patterns prompted the development of fractional RF (FRF) [2, 3].

FRF is often touted as being safe for all skin types due to its 'color-blind' nature. It should be noted that while RF interaction with skin is not affected by the presence of melanin, darker skin types and tanned skin are still susceptible to postinflammatory pigmentary changes. Since FRF exerts its tissue effects by heating and in most applications will induce some degree of wound healing response, the wise practitioner will exercise appropriate caution when treating high-risk skin.

Fractional Radiofrequency Technologies

FRF induces dermal heating, with or without epidermal ablation. Where ablation is induced, it is fractional, akin to fractional lasers, and therefore retains the advantages of rapid epithelialization and reduced downtime and lower risks of infection, pigmentary change, and scarring. FRF can be delivered in a noninvasive manner, applied directly to the skin surface, or in a minimally invasive manner, applied within the skin.

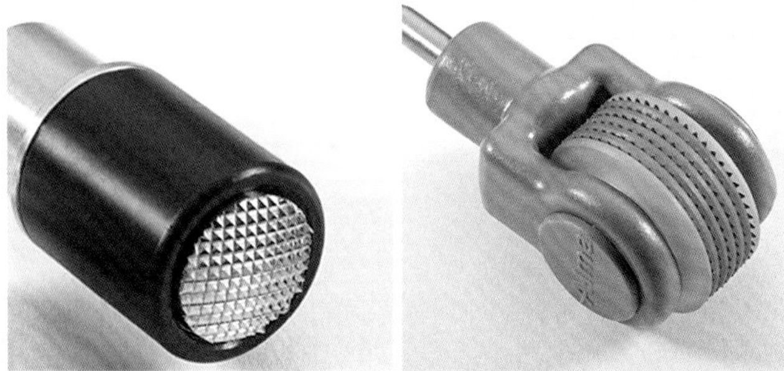

Fig. 2. Unipolar FRF: pixel RF stamping and rolling handpieces (Alma Lasers).

Each method has associated benefits and indications. A variant of FRF, called sublative RF, applies fractional bipolar RF to target the epidermis and superficial dermis using methods that require less energy to heat the collagen, in order to induce collagen contraction and neocollagenesis.

Noninvasive Fractional Radiofrequency

Noninvasive FRF can be delivered in several modalities: monopolar, bipolar, or a technology which applies bipolar RF in what has been termed sublative treatment. Since all three forms of RF require direct contact with an electrode, all three modalities are possible simply by fractionating the electrode that is in contact with the skin into a series of small electrodes. The geometry of the electrodes and the returns, however, determines the pattern in which the current travels. As a consequence, the main differentiator between these three approaches is the pattern of thermal damage induced in the tissue.

Unipolar FRF is applied with a single active electrode, whose contact with the skin is made with a series of pinpoint extensions from the handpiece electrode. The RF travels in proximity to each contact point on the tissue, with radial decrement in energy. An example of unipolar FRF is the Pixel RF (Alma Lasers, Israel). The device utilizes handpieces which have either a grid of pins or a rolling wheel with pins; these pins allow the transfer of current to the skin when placed in gentle, direct contact with the skin (fig. 2). The tissue effect of unipolar RF is very similar to that of lasers since the majority of the energy is absorbed at the point of contact, and heating decreases with distance from the electrode (fig. 3) [4]. The energy at the point of contact is highest, and may be so great as to induce heating to an ablative effect at the point of contact, with peripheral heating and coagulation in the surrounding tissue layers. A direct unipolar RF effect is predicated when there is direct contact between the electrode pins and the skin. The presence of gas, in particular nitrogen, at the skin surface, may allow generation of plasma in proximity to the electrode pins; the transfer of charge through the plasma may generate microsparks, adding to the fractional ablative effect.

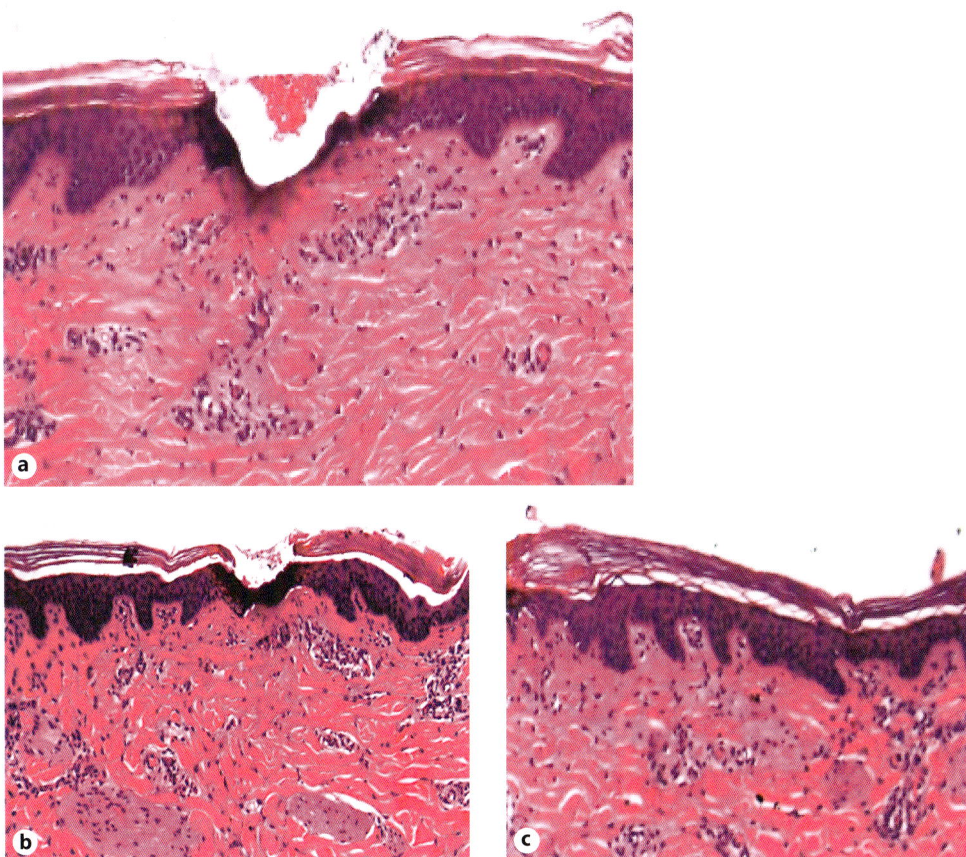

Fig. 3. Unipolar FRF tissue effects: day 0 (**a**), day 3 (**b**), and day 14 (**c**).

Bipolar RF is applied with a series of active and return electrodes (fig. 4). In straightforward bipolar RF, the current is limited to the tissue between the active and return electrodes. The distance between the electrodes affects the depth of penetration of the current. As a rough rule of thumb, the depth of penetration with marketed bipolar RF devices is one half the distance between the electrodes. The frequency of the RF generator also impacts the depth of penetration. Therefore, by varying the RF frequency and the distance between electrodes, the pattern of RF current can be modified to allow superficial or deeper heating when one is preferable to the other for a particular clinical indication (fig. 5).

Sublative RF is a variant of bipolar RF in which the epidermal effect is small and the dermal effect greater. Sublative RF is offered by the Matrix RF and eMatrix handpieces (Syneron-Candela, Israel). In contrast to fractional lasers and fractional unipolar RF, which induce a 'V'-shaped lesion in the skin, sublative RF generates a pyramid-shaped effect, by limiting the extent of heating in the uppermost layers and promoting the transfer of current in the deeper dermis (fig. 6). This effect is produced by stagger-

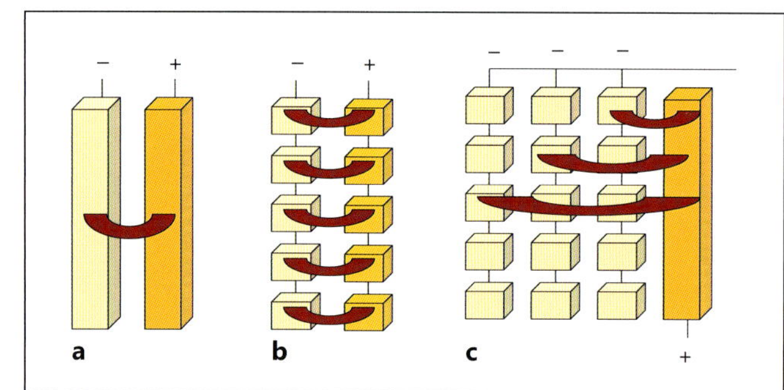

Fig. 4. Schematic of current flow in traditional bipolar RF (**a**), fractional bipolar RF (**b**), and sublative RF (**c**).

Fig. 5. Fractional bipolar RF handtips: Fractora (Invasix; **a**), FRF skin resurfacing (Endy Med; **b**), Infini (Lutronic; **c**).

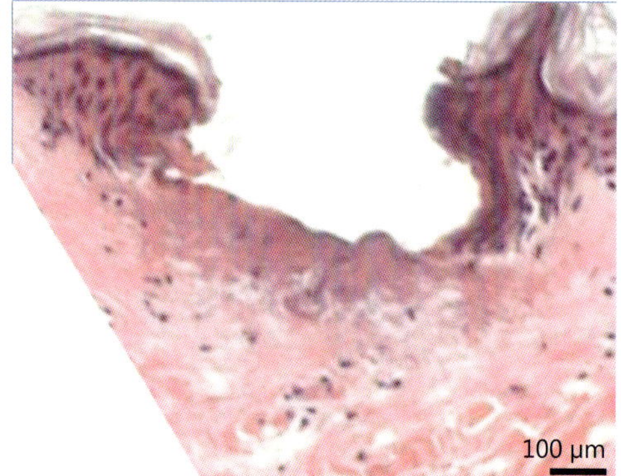

Fig. 6. Histology of sublative RF showing the trapezoidal pattern of heating, with a narrow zone of epidermal damage atop a broad zone of dermal coagulation (Matrix RF) [10].

ing the active electrodes in a matrix with variable distances from the return electrodes (fig. 7). This inverted geometry allows mild treatment of epidermal defects with a more aggressive treatment of the deeper dermis. Because the epidermal effect is localized to small foci, the approach allows a dermal treatment with more rapid epidermal healing and therefore shorter downtime and reduced risks of infection, pigmentary change, and scars.

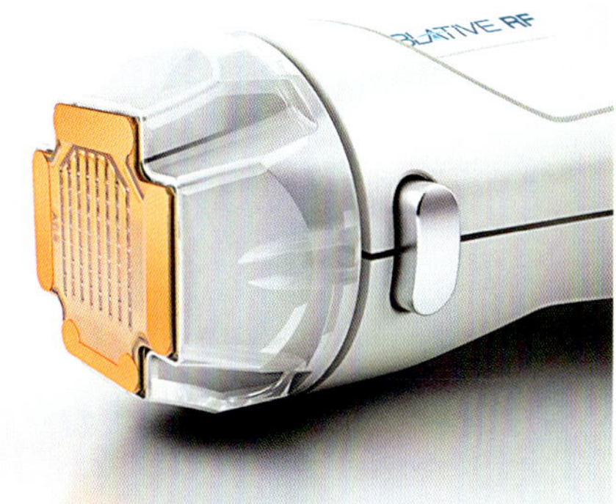

Fig. 7. Sublative RF tip formed by a matrix of electrodes with variable distance from a return electrode (Matrix RF and eMatrix).

Minimally Invasive Radiofrequency

Minimally invasive RF is discussed in detail in the chapter 'Minimally Invasive Radiofrequency'. A brief overview is included here to complement the FRF because minimally invasive RF is most often delivered in a fractional manner.

Minimally invasive RF is preferable in treatments where either tissue effects or safety warrant heating of the dermis and with no epidermal heating or ablation. This can be accomplished by direct delivery of RF in the dermis via microneedles. Several minimally invasive microneedle RF devices are marketed internationally at present. All provide needles of variable lengths and density (pins/cm^2). The devices vary by the use of insulation, which can be applied to the portion of the microneedle that is expected to be within the epidermis in order to prevent epidermal current transfer, and the application of real-time thermal monitoring. Some of these devices are cleared for hemostasis and cautery and are used off-label in aesthetic indications such as skin tightening and in treatments of wrinkles, cellulite, and acne scars. Since these treatments involve placement of multiple needles, treatments may require topical, regional, or tumescent anesthesia.

Representative microneedle device tips are shown in figure 8. These utilize an array of needles to introduce short pulses of RF energy into the dermis. The INTRAcel device (Jeysis, Korea) delivers bipolar and monopolar RF via insulated microneedles. Only the distal 0.3 mm is exposed to tissue to limit RF current to the dermis. The Infini (Lutronic, Korea) offers variable depth of the needles (0.5–3.5 mm), and the needles are placed sequentially, one row at a time, to reduce discomfort during insertion. The Infini needles are also insulated with the exception of the distal 0.3 mm. In contrast, the ScarLet (Viol, Korea), and the Secret RF (Danil SMC, Korea) utilize noninsulated needles, to provide heating throughout the needle tract. All four devices deliver the needles perpendicularly into the skin. These four have the benefit of covering

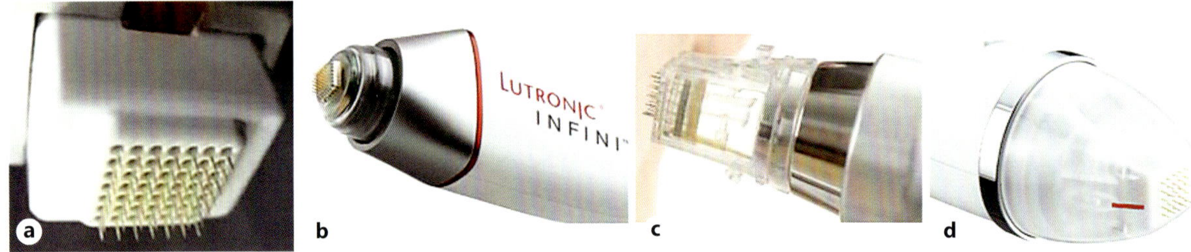

Fig. 8. Microneedle RF handpieces. **a** INTRAcel (Jeysis). **b** Infini (Lutronic). **c** ScarLet RF (Viol). **d** Secret RF (Danil SMC).

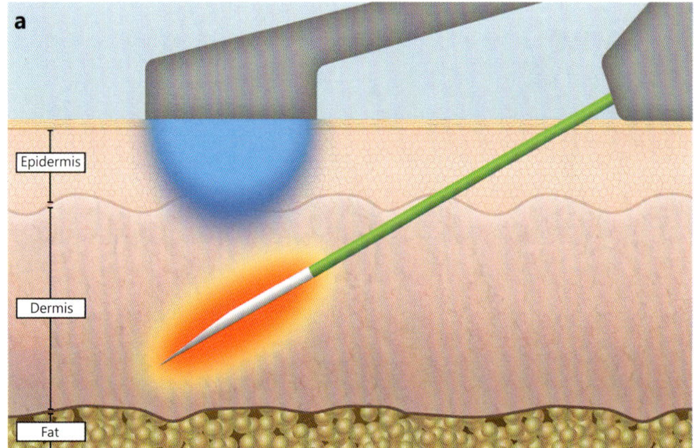

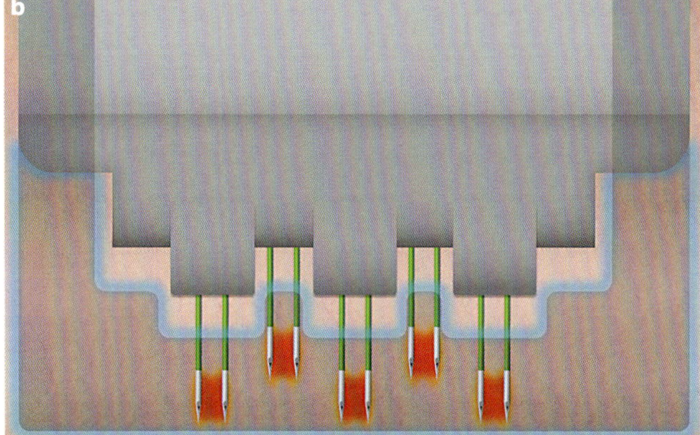

Fig. 9. Schematic of microneedle bipolar RF and single-use microneedle cartridge: ePrime (Syneron-Candela).

large areas quickly, as the handpieces have many needles and the current delivery time is rapid.

The ePrime microneedle RF (Syneron-Candela), marketed in the United States as Evolastin, applies RF current between five pairs of needles, where each pair of needles is an electrode pair (fig. 9). This approach limits the region of heating to

Halachmi · Onder · Fritz

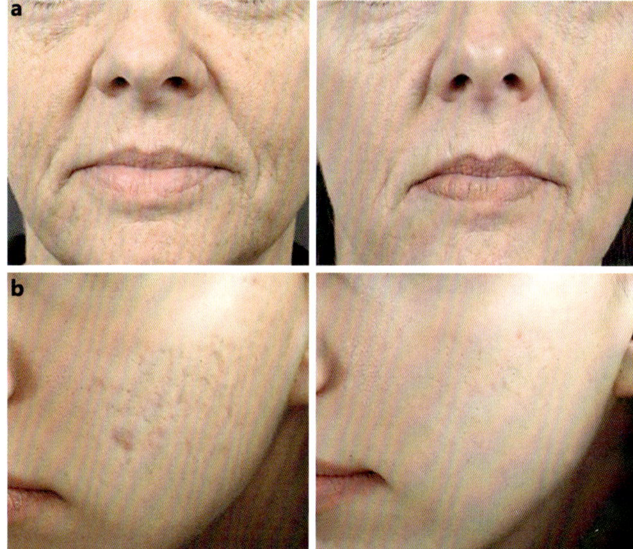

Fig. 10. Examples of response to FRF. **a** Photoaging before and after 2 treatments with sublative RF. Courtesy Dr. Sharyn Laughlin, Canada. **b** Acne scars before and after 3 treatments. Courtesy Dr. Amy Taub, USA.

the tissue between the needles. The needles are insulated and are inserted into the skin at a 30° angle to maximize horizontal coverage per insertion. A cooling tip is present on the handpiece at the point of epidermal contact to protect the epidermis further. The needles include a real-time feedback system which monitors tissue temperature. The operator inputs the desired dermal temperature (e.g. 65°C) and duration (e.g. 4 s); the device delivers RF energy until the temperature is reached and then maintains that temperature for the time entered. While treatment with the ePrime is longer than that of other microneedle devices, it offers greater control of tissue temperature and is likely able to achieve higher sustained temperatures than the rapid delivery devices. This dermal temperature control is not currently available with noninvasive or other minimally invasive FRF devices.

Representative responses to FRF are shown in figure 10.

Clinical Studies

Improvement of Photoaging and Wrinkles

Lee et al. [5] performed FRF treatments using a sublative FRF device (Matrix RF, Syneron-Candela) in 26 Korean female patients (mean age 56, skin phototype III–IV) with facial photodamage, using a 64-pin array handpiece. After 3 consecutive treatments performed at 4- to 6-week intervals, participants achieved more than 50% improvement in fine lines, pores, smoothness, tightness, brightness, and overall appearance. The FRF treatment was generally well tolerated, with mild edema lasting 36 h and mild erythema in some patients.

Bloom et al. [6] reported results after treating 25 females with mild to moderate wrinkles using a sublative RF device (eMatrix, Syneron-Candela). The study was performed with a 144-pin array using topical anesthesia. Three full-face procedures at 30-day intervals demonstrated improvement in skin laxity, texture, fine lines, and wrinkles. Adverse events were mild erythema and swelling. No post-inflammatory pigmentary changes were observed.

Dahan et al. [7] recently conducted a clinical trial with an FRF device that generates pulses of RF energy which are emitted into the skin, causing a nonablative deep dermal heating effect and resulting in skin tightening (EndyMed PRO, EndyMed Ltd., Israel). The study included 10 patients with Fitzpatrick wrinkle and elastosis scale of 5–8, who received a total of three treatments spaced one month apart. Results showed that all patients had a significant reduction in the Fitzpatrick wrinkle score. In all, 56% of patients reported no pain after treatment, while 44% reported minimal pain. Adverse events included transient erythema lasting up to 10 h after treatment.

FRF has also been assessed in darker skin types. Man et al. [8] evaluated the safety and efficacy of bipolar FRF (eMatrix) for the improvement of skin texture, fine lines, and wrinkles in 15 patients with Fitzpatrick skin types V–VI. All study participants received three full-face treatments spaced 30 days apart, using a 64-pin tip. Statistically significant improvements were noticed in most patients. No adverse events, particularly no postinflammatory hyperpigmentation or hypopigmentation, were seen. The study suggests that this FRF device can offer a safe and effective method of skin rejuvenation for skin types V through VI.

Acne Scars

There is literature on the use of fractional lasers in improving the cosmesis of acne scars. Like fractional laser, bipolar RF modalities are effective but require multiple treatment sessions to achieve a cosmetic improvement for this indication.

Gold et al. [9] performed a study evaluating the efficacy and tolerance of treatments in 15 patients with mild to moderate facial acne scars (Matrix RF). Treatments were performed with high energies (program C, 16–25 J, 1,000–1,500 μm) for maximal penetration [10–12]. Ten patients completed the study and found the fractional bipolar RF energy to be safe and effective in the treatment of acne scars.

Ramesh et al. [11] treated facial acne scars with the 'matrix-tunable radiofrequency' device in 30 patients with skin types ranging from IV to VI. Study participants had icepick, box, and rolling type scars of varying depths, sizes, and numbers present. Deep, box type scars of 8 patients were treated by subcision before RF treatment. Improvement at 2 and 6 months after the final treatment was assessed by visual comparison of baseline and posttreatment photographs. Results showed that the im-

   Halachmi · Onder · Fritz

provements were generally greater at 6 months (20–70%) than at 2 months (10–50%). Cosmetic results were evaluated as good to excellent in 73% of patients, and in 100% of patients whose scars were pretreated with subcision. The improvement in icepick scars was generally greater than with box type scars, while improvement in rolling type scars was variable. Adverse events were limited to transient edema at the treatment site, and a burning sensation and erythema skin for several days following treatment. A few patients reported mild scaling and crusting. The authors concluded that matrix-tunable RF technology was effective in improving icepick scars and rolling type scars, and effective for box type scars when combined with subcision before treatment.

Sadick et al. [13] performed a study with a microfractional resurfacing device (EndyMed) for the treatment of acne scars, evaluating the effect of the FRF skin resurfacing applicator. Results showed a significant reduction in the depth of wrinkles and acne scars at 4 weeks after treatment, with further improvement at the 3-month follow-up. Investigators concluded that simultaneous RF fractional microablation and volumetric deep dermal heating had a beneficial clinical effect on wrinkles and acne scars.

Taub and Garretson [14] conducted a prospective study in 20 subjects of skin types II–V. All received up to 5 sublative treatments at 4-week intervals in which fractional 915-nm diode laser with bipolar (nonfractional) RF was followed by sublative RF (Matrix IR). Results were evaluated using the Goodman Scar Scale. Acne scars improved significantly at one month after 3 treatments; improvement persisted at least 12 weeks after the 5th treatment. Adverse effects were limited to transient erythema and edema.

Peterson et al. [15] evaluated the efficacy of the same device with a similar protocol. Treatments were very well tolerated, with no postinflammatory hyperpigmentation. Scar scale scores were reduced from baseline by 72.3% (p < 0.001) by day 210, with investigator-rated changes in scarring, texture, and pigmentation improving by 68.2% (p < 0.001), 66.7% (p < 0.001), and 13.3% (p = 0.05), respectively. Subjective patient scores were mixed: although overall improvement scores increased by 60% over baseline (p = 0.02), satisfaction scores were not statistically significant.

Striae

Striae distensae are dermal scars characterized by a flattening and atrophy of the epidermis. The treatment of striae remains challenging for clinicians, as truly effective therapeutic options are few. FRF devices have shown to be useful in the treatment and improvement of striae distensae.

Brightman et al. [16] applied sublative RF at high energy levels to patients with abdominal striae. Treatments resulted in a decreased atrophic appearance and improvement in the cigarette paper-like appearance of striae, as well as improvements in the telangiectasia within striae rubra. It was found that the number of treatments depends on the width and degree of atrophy of striae.

Suh et al. [17] used a fractional unipolar microplasma RF system followed by 'impact' therapy (Legato, Alma Lasers) to enhance penetration of platelet-rich plasma in treatment of striae in 18 participants. After 4 sessions at 2-week intervals, the average width of the widest striae decreased by 0.5 mm (from 0.75 to 0.27 mm). At 2 months, 72% of participants reported 'good' or 'very good' improvement. Two patients (11%) reported hyperpigmentation. The authors conclude that the treatment is safe and effective.

Conclusion

The most common indications for fractional lasers include the reduction of wrinkles and skin laxity, and acne scarring. The transient side effects of treatment with fractional lasers are erythema and edema, which are both safe and tolerable in the degree in which they occur. The aesthetic effect of fractional bipolar RF-based devices is achieved through thermal damage of dermal collagen generated by an electric current. As a consequence, fractional bipolar RF energy stimulates wound healing, dermal remodeling, neoelastogenesis, neocollagenesis, and hyaluronic acid formation.

Fractional bipolar RF is an effective skin rejuvenation technology that has been shown to have a low side effect profile in patients with Fitzpatrick skin types I–IV. Because RF interaction with skin is not affected by the degree of melanin, it may be safer in some applications than lasers that operate in the visible light spectrum. However, postinflammatory changes can occur with RF as with all interventions; a conservative approach is the best accessory to any energy-based device.

References

1 Manstein D, Herron GS, Sink RK, Tanner H, Anderson RR: Fractional photothermolysis: a new concept for cutaneous remodeling using microscopic patterns of thermal injury. Lasers Surg Med 2004;34: 426–438.

2 Krueger N, Sadick N: New generation radiofrequency technology. Cutis 2013;91:39–46.

3 Lolis M, Goldberg DJ: Radiofrequency in cosmetic dermatology: a rereview. Dermatol Surg 2012;38: 1765–1776.

4 Halachmi S, Orenstein A, Meneghel T, Lapidoth M: A novel fractional micro-plasma radio-frequency technology for the treatment of facial scars and rhytids: a pilot study. J Cosmet Laser Ther 2010;12: 208–212.

5 Lee HS, Lee DH, Won CH, Chang HW, Kwon HH, Kim KH, Chung JH: Fractional rejuvenation using a novel bipolar radiofrequency system in Asian skin. Dermatol Surg 2011;37:1611–1619.

6 Bloom BS, Emer J, Goldberg DJ: Assessment of safety and efficacy of a bipolar fractional radiofrequency device in the treatment of photodamaged skin. J Cosmet Laser Ther 2012;14:208–211.

7 Dahan S, Rousseaux I, Cartier H: Multisource radiofrequency for fractional skin resurfacing-significant reduction of wrinkles. J Cosmet Laser Ther 2013;15: 91–97.

8 Man J, Goldberg DJ: Safety and efficacy of fractional bipolar radiofrequency treatment in Fitzpatrick skin types V–VI. J Cosmet Laser Ther 2012;14:179–183.

9 Gold MH, Biron JA: Treatment of acne scars by fractional bipolar radiofrequency energy. J Cosmet Laser Ther 2012;14:172–178.

10 Hruza G, Taub AF, Collier SL, Mulholland SR: Skin rejuvenation and wrinkle reduction using a fractional radiofrequency system. J Drugs Dermatol 2009;8: 259–265.

11 Ramesh M, Gopal M, Kumar S, Talwar A: Novel technology in the treatment of acne scars: the matrix-tunable radiofrequency technology. J Cutan Aesthet Surg 2010;3:97–101.

12 Yeung CK, Chan NP, Shek SY, Chan HH: Evaluation of combined fractional radiofrequency and fractional laser treatment for acne scars in Asians. Lasers Surg Med 2012;44:622–630.

13 Sadick NS, Sato M, Palmisano D, et al: In vitro animal histology and clinical evaluation of multisource fractional radiofrequency skin resurfacing (FSR) applicator. J Cosm Laser Ther 2011;13:204–209.

14 Taub AM, Garretson CB: Treatment of acne scars of skin types II to V by sublative fractional bipolar radiofrequency and bipolar radiofrequency combined with diode laser. J Clin Aesthet Dermatol 2011;4:18–27.

15 Peterson JD, Palm MD, Kiripolsky MG, Guiha IC, Goldman MP: Evaluation of the effect of fractional laser with radiofrequency and fractional radiofrequency on the improvement of acne scars. Dermatol Surg 2011;37:1260–1267.

16 Brightman L, Goldman MP, Taub AF: Sublative rejuvenation: experience with a new fractional radiofrequency system for skin rejuvenation and repair. J Drugs Dermatol 2009;8:9–13.

17 Suh DH, Lee SJ, Lee JH, Kim HJ, Shin MK, Song KY: Treatment of striae distensae combined enhanced penetration platelet-rich plasma and ultrasound after plasma fractional radiofrequency. J Cosmet Laser Ther 2012;14:272–276.

Shlomit Halachmi, MD, PhD
Herzelia Dermatolgy and Laser Center
9 Hamenofim St.
Herzelia Pituach (Israel)
E-Mail drshlomithalachmi@gmail.com

Lapidoth M, Halachmi S (eds): Radiofrequency in Cosmetic Dermatology.
Aesthet Dermatol. Basel, Karger, 2015, vol 2, pp 62–69 (DOI: 10.1159/000368735)

Minimally Invasive Radiofrequency

R. Stephen Mulholland[a] · Shlomit Halachmi[b]

[a]Private Plastic Surgery Practice, Toronto, Ont., Canada; [b]Herzelia Dermatology and Laser Center,
Herzelia Pituach, Israel

Abstract

Minimally invasive radiofrequency is most commonly applied in a fractional manner (fractional radiofrequency, or FRF). The approach relies on introduction of a set of fine needle electrodes into the skin, which are then activated to deliver energy. Its role is similar to that of fractional lasers, namely to induce neocollagenesis by thermal effects within the dermis. The fractional thermal injury of deep dermal collagen induces a vigorous wound healing process with dermal remodeling and the generation of new collagen, elastin and hyaluronic acid. The major benefit of minimally invasive FRF is that, depending on the needle configuration and insulation, energy can be deposited directly within the dermis with no thermal effects to the epidermis. Side effects are minimal and typically include transient erythema lasting approximately 2 days. It is a viable nonsurgical therapeutic option for the improvement of numerous cosmetic indications including skin rejuvenation, facial skin laxity, rhytides, acne scars, large pores, and photoaged skin.

Laser light and radiofrequency (RF) energy sources have succeeded in effectively treating the typical signs of skin aging. Nonsurgical devices have become very popular in the treatment of rhytides and skin laxity as patients prefer to achieve aesthetic improvement of these cosmetic indications with minimal side effects and downtime.

A novel 'minimally invasive RF' approach employing a bipolar microneedle electrode system was recently introduced and its thermal effects on human skin in vivo were characterized in some studies.

To improve the efficacy and predictability of nonsurgical treatments, a minimally invasive bipolar RF system was developed with real-time tissue temperature feedback that delivers fractional RF (FRF) thermal pulses to the reticular dermis. This system safely creates thermal injury zones in a fractional manner within the reticular dermis without damage to the dermal-epidermal junction, epidermis or subcutaneous tissue.

Adnexal structures such as hair follicles, and sweat and sebaceous glands are spared during FRF treatment. Histological evidence of neocollagenesis and neoelastogenesis following treatment suggests the replacement of dermal volume.

The currently marketed devices can be categorized into various platforms. These will be described below.

ePrime Platform

The e-Prime system was the first to be reported. It is based on the work of Hantash et al. [1]. The system uses a handpiece with detachable cartridges equipped with multiple pairs of needles. The 32-gauge needles penetrate the skin at a 25° angle. The procedure, which is performed under local anesthesia, utilizes RF to produce thermal heating to stimulate neocollagenesis and neoelastogenesis. Hantash et al. [1] first reported a 35-watt bipolar RF device then called (Renesis, Primaeva Medical Inc., Fremont, Calif., USA), configured with 5 microneedle pairs of 30-gauge electrodes. The device is currently marketed as ePrime or Evolastin.

In this first human pilot study, 15 adult patients undergoing elective surgical facelift or abdominoplasty were evaluated. All patients were treated with local anesthesia except for one patient who underwent general anesthesia just prior to treatment. Tissue impedance, power and temperature were measured. Histological examination revealed neocollagenesis, elastin replacement, and heat shock protein response. Results showed that a minimally invasive bipolar microneedle RF device can create thermal zones within the reticular dermis, and that lesion size can be controlled by treatment pulse and lesion temperature.

Alexiades-Armenakas et al. [4] recently performed a study using a minimally invasive bipolar FRF system (ePrime™, Syneron Medical Ltd., Yokneam, Israel). This prospective, open-label, multicenter clinical trial enrolled 100 patients with mild to severe facial and neck rhytides who received a single treatment with the device. Treatment parameters were target temperature, treatment duration and number of insertions. Results showed that 96% of patients demonstrated improvement of rhytides at the 3-month follow-up (after a single treatment), which improved to 100% at the 6-month follow-up. The data indicate that a substantial improvement can be achieved by 3 months after treatment, with continued improvement to at least 6 months.

In another clinical study conducted by Alexiades-Armenakas et al. [5], 15 patients with facial skin laxity received FRF treatments with Miratone™ system (Primaeva Medical Inc., Pleasanton, Calif., USA), a minimally invasive bipolar FRF device. Baseline and follow-up digital photographs of the 15 patients were randomly mixed with 6 sets of baseline and follow-up images of patients who underwent surgical facelift with equivalent baseline facial laxity grades, and all images were compared and graded by 5 independent blinded evaluators. The side effects of patients who underwent

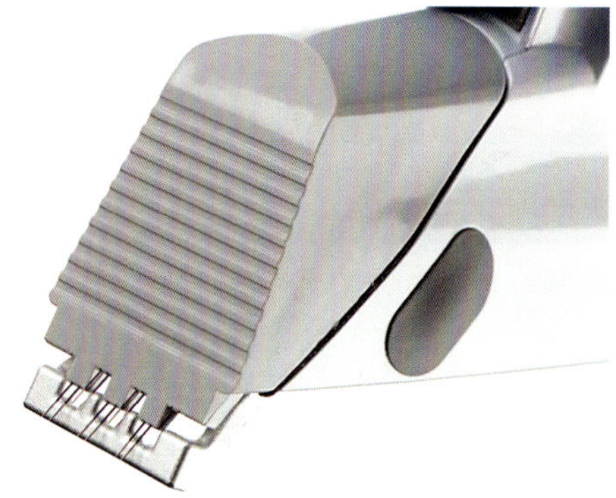

Fig. 1. Evolastin (ePrime) cartridge.

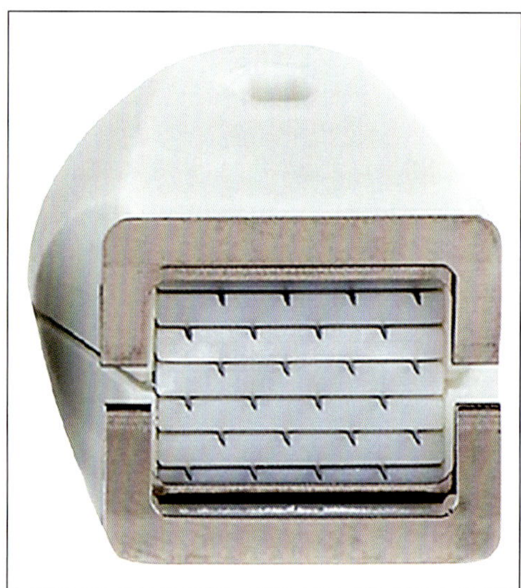

Fig. 2. Fractora tip.

FRF treatment included transient erythema, mild edema and mild to moderate purpura that resolved within 5–10 days. Results showed that the patients who underwent a surgical facelift had a 49% improvement in skin laxity over baseline, compared with 16% of patients treated with FRF. Minimally invasive FRF had an effect in skin laxity without adverse effects and complications of surgical procedure, suggesting that the minimally invasive FRF treatment could provide an important nonsurgical treatment option for the treatment of facial skin laxity [5].

The use of minimally invasive microneedle electrode pairs and bipolar energy has several advantages. It is a very direct method to control the precise location within the skin where energy is delivered, and allows tissue characteristics to be measured. When the microelectrodes were intentionally deployed subcutaneously into the adipose tissue, the impedance was significantly higher than the reticular dermis. When elec-

Mulholland · Halachmi

trodes were deployed in the papillary dermis, impedance was lower than the reticular dermis. A direct benefit of the minimally invasive approach is the ability to obtain real-time feedback of treatment effects through the use of temperature sensors at the distal end of the microelectrode.

As this treatment is a minimally invasive procedure, local anesthesia proved sufficient to achieve patient comfort. Manufacturers of noninvasive monopolar or bipolar RF devices recommend against infiltration with local anesthetics as they may alter energy deposition and heating patterns. The use of local anesthetics provides an advantage in pain management [1].

Fractora Platform

The Fractora (Invasix, Yokeam, Israel) FRF resurfacing system utilizes a reusable handpiece, into which a Fractora disposable tip is inserted. Each tip consists of a variable density and variable-length sharp array of needles, all of which are positively charged needle electrodes. On the sides of each Fractora tip are flat, negatively charged electrodes which receive the RF energy from the positively charged needle electrodes. RF energy is generated in the console (Inmode or Bodytite, Invasix, Yokeanam, Israel), that then travels down the electrical cord into the reusable Fractora handpiece. The RF energy from the needle-electrode creates an ablative crater, surrounded by a circular ring of nonablative, but irreversible, coagulative thermal injury and finally a wide, broad, deep zone of nonablative RF thermal stimulation as the current flows from the tip of the positively charged pin to the negatively charged side electrodes.

There are a variety of Fractora which can be categorized by tip density, with a high-density or low-density epidermal impact, or by depth of the needle and hence ablative injury and RF penetration, from 600 to 3,000 μm. Based on numerous abdominoplasty histologies, the ablative depth and clinical wound healing profiles have been well documented, and the clinical protocols are based upon these biological/histological studies, and then multicenter empirical skin trials [6].

(a) Nonablative tip. High-density, shallow dermal depth impact tip: for almost a purely nonablative effect and very little epidermal impact. This tip has a 60-pin array, and the ablative craters are 150–300 μm in depth.

(b) 60-pin, 600-μm-depth tip. Moderate epidermal impact, with a density array with 5–8% epidermal coverage (without overlap). Mid-dermal vertical impact tip: 60 pins for moderate density and a 600-μm-depth impact (mid-papillary dermal effect).

(c) 126-pin, 600-μm-depth tip. Higher epidermal impact with 10–16% coverage (without overlap) and a mid-dermal depth impact of 600 μm.

(d) 24-pin, uncoated 3,000-μm-depth tip. Lower density, lower epidermal impact (less that 4% coverage without overlap), but transdermal 3,000-μm-length needles for a very deep dermal impact.

(e) 24-pin, silicon-coated, 3,000-μm-depth tip. This tip, also called the BiFrax tip (for Bifractional, horizontal and vertical fractional injury) is designed for selective deep thermal ablation, but with no epidermal-thermal impact (because of the proximal silicone coating) and thus has worked well for treating lighter skin patients who want a 'lift' effect or tightening, with minimal superficial effects.

The Fractora can be deployed in the same fashion as fractional CO_2, or fractional erbium laser resurfacing in skin types I, II and III. The treatment technique is similar with all tips; specifically, one can deliver a pulse every second to 2.5 pulses per second. Overlap is approximately 30–50%, with care taken not to overlap one negative electrode on top of another, or there will be too much superficial positive current flowing up to the negative electrode in that area, and there will be a strip of epidermal lysis. With 30–50% overlap, the aesthetic physician or technician moves across the entire treatment area.

With RF energy fluences between 10 and 25 mJ per pin, the clinical ablative effect is much like that seen with infrared fractional microthermal zone injuries. These lower-fluence Fractora treatments can be performed with simple topical anesthesia without nerve blockade or diffuse hypodermal infiltration. A Zimmer cyro air chiller device is highly recommended for all ablative treatments, laser or RF.

At all energy levels with the Fractora, when treating the face, there is an RF-mediated stimulation of the facial nerve, as RF travels half the distance between the positive pin electrodes and the negative side electrodes, which will stimulate the facial nerve and result in contraction of some of the facial muscles. Although at moderate energies patients do not find this contraction painful, they should be warned as it is an unusual sensation that they will not have experienced with lasers. With higher energies, the facial contractions can be stronger, and using your contralateral hand to flatten the skin and facial muscles will greatly minimize their contractile movement. In general, with high fluence (those over 40 mJ/pin) I will use hypodermal infiltration with double-strength Klein solution, with or without V1, V2 and V3 nerve blocks to ensure a comfortable treatment. With this hypodermal 'wetting' aesthesia, high-energy Fractora patients are comfortable at all energy levels. This hypodermal 'wetting' is very simple to perform, is not as distensive as tumescent lipo-infiltration (and, by the way, I must also perform this type of anesthesia for fractional CO_2 resurfacing, at higher settings).

With Fractora FRF energies between 25 and 40 mJ per pin, the ablative effect is like a mild, low-fluence CO_2 or more aggressive erbium treatment. At these energies, topical anesthesia, a Zimmer air cooler and possibly nerve blocks will be required to make the treatments comfortable. Between 40 and 62 mJ per pin, these high-fluence ablative energies are very similar in pain and discomfort to a strong fractional CO_2 treatment, and topical anesthetic alone is not recommended. A Zimmer cooler and nerve blocks, with or without hypodermal dilute anesthetic 'wetting' infiltration is required. There are various protocols that can be used with the Fractora FRF resurfacing. With mild energies, a treatment can be performed every week to 3 weeks, and a series of 3–6

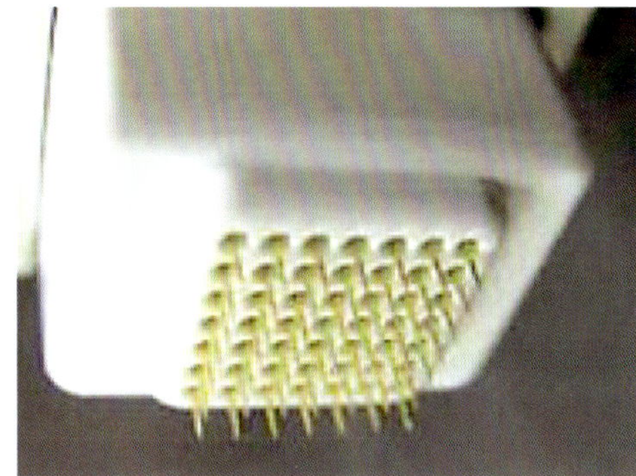

Fig. 3. INTRAcel tip.

treatments is required to achieve results similar to those seen with multiple low-level fractional CO_2, erbium, or infrared fractionals. However, the FRF resurfacing device can also be used in high-fluence, single-treatment protocols. High-fluence, single treatment protocols would deploy the clinical tip most relevant to the pathology presented by the patient.

INTRAcel Platform

Cho et al. [7] recently performed a clinical study using a minimally invasive FRF microneedle device (INTRAcel, Jeisys, Seoul, South Korea) in 30 patients with acne scars and large facial pores. This bipolar RF technology delivers its RF energy through 49-microneedle electrodes in an area of 1 cm^2, which are deployed into the deep dermis perpendicular to the skin surface, creating multiple partial thermal injury columns in the target area. The entire needle electrode is nonconductive except for the tip (beginning at 0.3 mm from the distal end), thereby protecting from RF heating and thermal tissue damage at the insertion site. The currently available microneedles are 0.5, 0.8, 1.5, or 2.0 mm long, and the RF energy delivery durations differ according to set energy levels. In this study, only a 1.5-mm needle at 500-watt power (maximum power 700 W) was employed. Prior to treatment, all makeup was removed, the face was cleaned with a facial foam cleanser and 70% alcohol, and a topical lidocaine-prilocaine anesthetic cream was applied under occlusion to both cheeks for 30–60 min. A full-face, double-pass treatment was then performed without the simultaneous use of an epidermal cooling device. Results were very positive, with physician evaluation and patient satisfaction showing an overall improvement in acne scars and large pores in more than 70% of all patients. The investigators concluded that clinical improvement appeared to be related to dermal matrix regeneration, and that the FRF microneedle treatment may be effective in improving acne scars and facial pores [7].

Fig. 4. Scarlet RF tip.

Scarlet Platform

The Scarlet™ (Viol Co., South Korea) adopted noninsulated microneedle electrodes. This device utilizes disposable single-use treatment tips consisting of 5 noninsulated microneedle electrode pairs per 10 mm² area, with the exposed electrode extending from 0.5 to 3 mm below the surface of the skin. Having an outer diameter of 0.3 mm each, these bipolar electrode pins form a closed circuit through the irradiated skin, delivering 2 MHz of conducted RF current to the skin. It appears the needles were designed to be noninsulated to reduce bleeding and to broaden the electric treatment field in the dermis. Seo et al. [3] found the microneedle FRF device to be a safe and effective treatment for skin rejuvenation in Asian patients. The microneedle RF treatment combined with a stem cell-conditioned medium showed a synergistic effect for skin rejuvenation.

In another study, Seo et al. [8] used a microneedle FRF device (Scarlet, Viol Co.) in 25 female patients, evaluating its safety and efficacy for the treatment of skin rejuvenation. In this study, patients received a total of 3 FRF treatments spaced 4 weeks apart. Adverse events were minimal and limited to mild pain and transient erythema during and after the procedures. In general, the incidence of side effects is lower with FRF compared to other modalities such as nonablative fractionated lasers. The authors concluded that microneedle fractional bipolar RF treatment is effective for the treatment of photoaged Asian faces with darker skin types, and has little risk of postinflammatory hyperpigmentation. The main limitations of this study were the small number of patients and lack of long-term follow-up after the final treatment [8].

Conclusion

The results of the above-mentioned studies suggest that 'minimally invasive FRF treatment' is a viable nonsurgical therapeutic option. The fractional thermal injury of deep dermal collagen induces a vigorous wound healing process with dermal remodeling and the generation of new collagen, elastin and hyaluronic acid. All studies recommend the application of local anesthesia prior to treatment. Side effects are minimal and typically include transient erythema lasting approximately 2 days. Following treatment, patients should avoid makeup for 24 h, and minimize sun exposure for approximately 14 days. These studies suggest that minimally invasive FRF treatment may provide an important nonsurgical alternative for the improvement of facial skin laxity, rhytides and acne scars.

References

1 Hantash BM, Renton B, Berkowitz RL, Stridde BC, Newman J: Pilot clinical study of a novel minimally invasive bipolar micro needle radiofrequency device. Lasers Surg Med 2009;41:87–95.

2 Taub Forman A: Evolastin: a minimally invasive procedure for volume, lifting, and tightening. Pract Dermatol April 2012.

3 Seo KY, Kim DH, Lee SE, Yoon MS, Lee HJ: Skin rejuvenation by microneedle fractional radiofrequency and a human stem cell conditioned medium in Asian skin: a randomized controlled investigator blinded split-face study. J Cosmet Laser Ther 2013; 15:25–33.

4 Alexiades-Armenakas M, Newman J, Willey A, Kilmer S, et al: Prospective multicenter clinical trial of a minimally invasive temperature-controlled bipolar fractional radiofrequency system for rhytid and laxity treatment. Dermatology Surgery 2013;39: 263–273.

5 Alexiades-Armenakas M, Rosenberg D, Renton B, et al: Blinded, randomized, quantitative grading comparison of minimally invasive, fractional radiofrequency and surgical face-lift to treat skin laxity. Arch Dermatol 2010;146:396–405.

6 DiBernado BE: Randomized, blinded, split abdomen study evaluating skin shrinkage and skin tightening in laser-assisted liposuction versus liposuction control. Aesth Surg J 2010;30:593–602.

7 Cho SI, Chung BY, Choi MG, Baek JH, Cho HJ, Park CW, Lee CH, Kim HO: Evaluation of the clinical efficacy of fractional radiofrequency micro needle treatment in acne scars and large facial pores. Dermatol Surg 2012;38:1017–1024.

8 Seo KY, Yoon MS, Kim DH, Lee HJ: Skin rejuvenation by micro needle fractional radiofrequency treatment in Asian skin; clinical and histological analysis. Lasers Surg Med 2012;44:631–636.

R. Stephen Mulholland
Private Plastic Surgery Practice
66 Avenue Road Suite 4
Toronto, ON M5R3N8 (Canada)
E-Mail mulhollandmd@spamedica.com

Lapidoth M, Halachmi S (eds): Radiofrequency in Cosmetic Dermatology.
Aesthet Dermatol. Basel, Karger, 2015, vol 2, pp 70–80 (DOI: 10.1159/000368736)

Radiofrequency Combinations

Shimon Eckhouse[a] · Klaus Fritz[b, c] · Maurice A. Adatto[d]

[a]Syneron-Candela, Yokneam, Israel; [b]Dermatology and Laser Centers, Landau, Germany; [c]Carol Davila
University, Bucharest, Romania; [d]Skinpulse Dermatology and Laser Center, Geneva, Switzerland

Abstract

Radiofrequency (RF) may be combined with other modalities, particularly light-based treatments, to
maximize the benefits of each while reducing the risks of treatment. The light energy may be used,
for example, to preheat a target by selective photothermolysis. Since RF travels preferentially in
warmer tissues due to reduced impedance, the combination of RF and light may allow the use of
lower fluence of each. Such an approach can maintain effectiveness and improve safety. This approach has been used successfully on a variety of indications on the face and body. This chapter
reviews the clinical studies reported. © 2015 S. Karger AG, Basel

Radiofrequency (RF)-based devices have become a mainstay in modern aesthetic
medicine and are considered to be one of the leading technologies currently available for cosmetic treatments. Their proven safety and efficacy has been extensively
documented in troves of clinical trials for the treatment of a plethora cosmetic indications that include skin rejuvenation, skin tightening, wrinkle improvement, acne
vulgaris, acne scarring, cellulite, striae distensae, hair removal and vascular lesions.
Varying combinations of RF energy with other therapeutic modalities such as laser
and light energy sources as well as mechanical tissue manipulation with vacuum and
massage have also been investigated and explored in depth, and have been shown to
even further improve the clinical outcomes of these cosmetic indications.

elōs (electro-optical synergy) (Syneron Medical Ltd., Yokneam, Israel) was the
first technology that simultaneously harnessed the power of bipolar RF and optical
energy, resulting in a synergistic effect of combined technologies. The technology
overcomes some of the safety and procedural limitations of intense pulse light (IPL)
and conventional lasers.

Some of the most widely used combination RF systems are those that use IPL, a diode laser, or infrared (IR) light. Several different devices based on the elōs technology concept as well as other technologies and approaches have been shown to successfully address many different cosmetic indications safely and effectively.

Studies and Evidence-Based Efficacy

Skin Rejuvenation

One combination RF system (Aurora SR, Syneron Medical Ltd., Yokneam) uses IPL as its optical energy source, with wavelengths between 400 and 980, 580 and 980, and 680 and 980 for different targets or chromophores [1]. RF energies up to 25 J/cm^3 can be generated with a dermal penetration of 4 mm [2].

El Domyati et al. [3] performed a clinical trial using the Aurora™ device in 6 patients with Fitzpatrick skin types 3–4 and Glogau class I–II wrinkles. Study participants received 6 treatments performed at 2-week intervals in the periorbital region, and punch biopsies were taken to evaluate the histological changes following treatment. Histologic analysis revealed an increase in epidermal thickness, a 53% reduction in elastin content and 28% increase in collagen fibers. The authors found that the treatment modality could stimulate the repair processes and reverse the clinical as well as the histopathological signs of skin aging. They concluded that electro-optical synergy is an effective treatment for contouring facial skin laxity, while having the advantage of being a risk-free procedure with minimal recovery time [3].

Sadick et al. [4] reported similar positive results in 108 consecutive patients who received a series of full-face treatments with the Aurora™ system. In this clinical trial, patients received 5 full-face treatments every 3 weeks, with each treatment consisting of 1–8 full-face and segmental passes. The number of passes, specific wavelength of pulsed optical energy, and RF energy were determined by the patient's skin type, dyschromia, wrinkle pathology, and presence of tan. Results of the study showed an overall skin improvement of 75.3%. Overall average wrinkle improvement was 41.2%, pore size and pigmentation improved by 65%, improvement in skin laxity was rated at 62.9% and skin texture was reported to improve 74.1%. Overall patient satisfaction was 92%. The authors concluded that the device was a safe and effective treatment modality for noninvasive skin rejuvenation [4].

Facial and Neck Wrinkles

Another combination RF system (Polaris WR, Syneron Medical, Ltd.) uses a combined 900-nm diode laser with RF energy for the treatment of deep wrinkles and the superficial signs of photoaging. The optical and RF energies in the device are delivered simultaneously through a bipolar electrode tip, with optical energy fluences ranging from 10 to 50 J/cm^2 and RF energies from 10 to 100 J/cm^3 [1].

Doshi and Alster [5] conducted a clinical trial using the Polaris device, evaluating the safety and efficacy of the combination RF/diode laser device for the treatment of skin laxity and facial rhytides. The study included 20 female patients (skin phototypes I–III) with mild to moderate rhytides and skin laxity who received a total of 3 treatments performed in 3-week intervals. Results showed that the majority of patients could achieve a modest improvement in facial rhytides 6 months after treatment, with a greater improvement in periorbital than perioral wrinkles. The investigators found the device to be a safe and effective treatment modality for the improvement of mild to moderate facial rhytides and skin laxity.

Hammes et al. [6] also performed a study with the Polaris device, evaluating the safety and efficacy of treatment in 24 patients with periorbital and perioral wrinkles. Study participants received 6 treatments spaced 4 weeks apart. Each treatment consisted of 2 passes with the device using a maximum RF energy of 100 J/cm^3, and a maximum laser energy of 50 J/cm^2. Three months after the last treatment session, results showed that patients could achieve a moderate improvement of their facial wrinkles, and no difference was noted between periorbital and perioral wrinkle reduction. A notable wrinkle reduction was recorded in 58% of patients, and 16% of patients noted a mild to moderate transient edema and erythema lasting up to 24 h. The authors concluded that the Polaris device was effective in the reduction of perioral and periorbital wrinkles, and suggested that the clinical effect may be caused by matrix coagulation of the deep dermis by the RF and selective thermolysis of blood vessels with the 900-nm diode laser.

In a study performed by Sadick et al. [7], the Polaris device was evaluated for its safety and efficacy in the treatment of wrinkles and skin texture. The multicenter clinical trial included 30 patients with Glogau grade II–IV facial wrinkles who received up to 3 treatment sessions with the device spaced 2–3 week apart. Study participants were treated with a fluence of 30–50 J/cm^2 of diode laser energy and RF energy of 80–100 J/cm^3. In the 23 patients who completed all 3 treatment sessions (7 patients were lost to follow-up), results showed that more than 50% had a greater than 50% improvement in the appearance of wrinkles, and all participants reported a noticeable improvement in skin smoothness and texture. The authors concluded that the combination of diode laser and RF energies decreases the appearance of wrinkles and improves skin texture.

Striae Distensae

Striae distensae are dermal scars characterized by a flattening and atrophy of the epidermis. The treatment of striae distensae is challenging and although many different therapeutic avenues have been used including the application of topical creams, local mechanical stimulation as well as energy-based modalities, a gold-standard treatment approach still remains elusive. Lately, RF energy in combination with other energy-based technologies has been used to improve the appearance of striae distensae.

Ryu et al. [8] conducted a study to evaluate the efficacy and safety of a combination therapy with fractionated microneedle RF and fractional CO_2 laser in the treatment of striae distensae. The study included 30 female patients with moderate to severe striae distensae who were evenly divided into three treatment groups: fractional CO_2 laser only (n = 10), microneedle RF only (n = 10), and combination therapy (n = 10). Improvement was evaluated using a visual analogue scale (range 1–4). Results showed that the mean clinical improvement score of the dermatologist was 2.2 in the fractional CO_2 laser-treated group, 1.8 in the microneedle RF-treated group, and 3.4 in the combination treatment group. Histologic examination showed a thickened epidermis and a clear increase in the number of collagen fibers in the microneedle RF- and fractional CO_2 combination-treated sites. The investigators concluded that the combination therapy of fractionated microneedle RF and fractional CO_2 laser is a safe treatment approach with a positive therapeutic effect on striae distensae [8].

The Legato™ system (Alma Lasers Ltd., Israel) is a novel technology that employs both RF energy and ultrasound to facilitate the transepidermal delivery of cosmeceutical products, allowing for a deeper penetration into the targeted tissue. This innovative system has been found to be useful in improvement of striae distensae as well as scars and skin rejuvenation.

Issa et al. [9] recently used the Legato system to evaluate the safety and efficacy of transepidermal delivery of retinoic acid 0.05% cream using ablative fractional RF and acoustic pressure wave ultrasound technology in 8 patients with alba type striae distensae on the breast. Each of the patients received 4 treatments with the Legato system, while another 8 patients with alba type striae distensae on the abdominal area received RF treatment alone, without the application of retinoic acid or ultrasonic energy. Results showed that a significant improvement could be achieved in the appearance of striae distensae in all patients treated with RF associated with retinoic acid 0.05% cream and ultrasound, with low incidence of side effects and a high level of patient satisfaction. Of the patients treated with RF alone, no significant improvements in the appearance of striae distensae were seen, and all patients recorded a low incidence of side effects as well as low patient satisfaction.

In a similar study, Suh et al. [10] also used the Legato system to evaluate the efficacy and safety of enhanced penetration of platelet-rich plasma with ultrasound after plasma fractional RF for the treatment of striae distensae on the abdomen. The clinical trial included 18 patients who received 4 treatment sessions spaced 2 weeks apart. Clinical outcomes were very positive, with 71.9% of patients reporting a 'good' or 'very good' overall improvement in the objective assessment, and 72.2% of patients reporting 'very satisfied' or 'extremely satisfied' with overall improvement in the subjective assessment. The investigators concluded that the fractional ablative RF and transepidermal delivery of platelet-rich plasma using ultrasound is useful in the treatment of striae distensae.

Cellulite

Typically characterized by an orange peel or cottage cheese-type dimpling of the skin, cellulite is a notoriously challenging to treat condition in which the affected skin appears to have areas with underlying fat deposits resulting in a dimpled, lumpy appearance. Most commonly occurring on the thighs and buttocks, cellulite affects approximately 85–98% of postpubertal females of all races, and while it is not a pathologic condition, it remains an issue of cosmetic concern to a great number of individuals. Although many therapeutic approaches have been tried including the attenuation of aggravating factors, pharmacological-based creams, endermologie, liposuction, subcision, phosphatidylcholine injections as well as laser and light-based technologies with and without additional tissue manipulation using vacuum and massage technologies, a truly effective gold-standard therapeutic approach remains elusive [11].

The VelaSmooth™ and VelaShape™ devices (Syneron Medical Ltd., Irvine, Calif., USA) are body contouring devices and part of Syneron's Vela line of body shaping products. They are designed to reduce or shrink the size of the fat cells and fat chambers so that cellulite becomes less noticeable and the skin smoother and more toned. The devices incorporate a combination of bipolar RF energy, IR light, as well as vacuum and mechanical massage, the combination of which has been clinically proven to be effective in circumferential and cellulite reduction. Simultaneously delivering 20 W of RF energy and IR light over a waveband ranging from 700 to 2,000 nm to the target area, mechanical massage is also given and suction is applied corresponding to 750 mm Hg of negative pressure. The effects of the vacuum and mechanical massage, combined with the delivery of heat energy, result in the metabolism of stored body fat and increase in lymphatic drainage.

In one study, Romero et al. [12] used the VelaSmooth device for the treatment of cellulite in 10 patients. Improvements in the overall appearance of cellulite and skin condition were achieved. The pain during treatment was reported as being bearable by all patients (no patients scored more than 5 points on an 11-point visual analogue scale) and the erythema lasted a few hours.

Sadick and Mulholland [13] treated 35 patients with cellulite twice weekly using the same device. After 16 treatments, skin smoothing, cellulite improvement and reductions in thigh circumference were seen.

In another study conducted by Wanitphakdeedecha and Manuskiatti [14], the VelaSmooth device was used in 12 women with cellulite who received treatments twice weekly for a total of 8–9 treatments. Results showed a beneficial effect on the reduction of abdomen and thigh circumference as well as a smoothening of cellulite.

Alster and Tanzi [15] performed a clinical trial with the VelaSmooth device in 20 patients with cellulite. Results showed that a 50% improvement in the appearance of cellulite could be achieved, and a 90% patient satisfaction.

Hexsel et al. [16] conducted a clinical trial using the VelaShape device in 9 patients who presented with body mass index (BMI) from 18 to 25 and at least a grade 6 on the Cellulite Severity Scale. Study participants received 12 treatments with the device on the posterior thighs and buttocks. Although a significant reduction of hip circumference could be achieved, no changes were seen in thigh circumference. Results also showed that the CSS specifically improved on both buttocks; however, no changes were observed on the thighs. The investigators concluded that the device is effective in the reduction of cellulite severity and body circumference measurements in the buttocks. In another study, Adatto et al. [17] looked at adipose tissue reduction and skin tightening using the Velashape II on 35 female patients who received one treatment per week over 6 weeks to their abdomen/flank, buttock, or thigh areas. Follow-up was up to 3 months post completion of the treatment protocol. Patient circumferences were measured and comparisons of baseline and post treatment outcomes were made. Diagnostic ultrasound (US) measurements were performed in 12 patients to evaluate the reduction in adipose tissue volume, and a cutometer device was used to assess improvements in skin tightening. A gradual decline in patient circumferences was observed from baseline to post six treatments. The overall body shaping effect was accompanied with improvement in skin tightening and was clearly noticeable in the comparison of the before and after treatment clinical photographs. These findings correlated with measurements of adipose tissue volume and skin firmness/elasticity using diagnostic US and cutometer, respectively. The thickness of the fat layer showed on average a 29% reduction between baseline and the 1-month follow up. The average reduction in the circumference of the abdomen/flanks, buttocks, and thighs from baseline to the 3-month follow-up was 1.4, 0.5, and 1.2 cm, respectively, and 93% of study participants demonstrated a 1-60% change in fat layer thickness. The authors concluded that, this new version with higher power especially in RF, is an effective modality for the reduction in circumferences of the abdomen/flank, buttock and thigh regions, and the improvement of skin appearance. 97% of the treated subjects were satisfied with the results at the follow-up visit.

ThermaLipo™ (Thermamedic Ltd., Alicante, Spain) is a bipolar RF device that uses a sensor that constantly monitors the return current as an indicator of impedance and adjusts the frequency to achieve total tissue volume penetration, while maintaining deeper tissue temperatures to ensure electrothermal damage deposition. The pulsed-mode RF has the ability to rapidly deposit a high-energy load, which increasingly raises the temperature of the subcutaneous tissue and skin. Van der Lugt et al. [18] conducted a study using the device in 50 patients with cellulite. Results showed a clear overall improvement of the shape of the buttocks, and less dimpled appearance was seen.

Skin Tightening

Winter [19] conducted a clinical trial using the VelaShape (Syneron Medical Ltd., Yokneam) and evaluated the efficacy of the device for reshaping and improving the skin texture and skin laxity in postpartum women. In the study, 20 patients received 5 treatments spaced one week apart to the abdomen, buttocks and thighs with the VelaShape system. Results showed that the trend for circumferential reductions continued and peaked at the 4-week follow-up, with more than 2-cm reductions seen in most treated areas. In addition, significant improvements in skin laxity were observed, and all treatments were well tolerated with no major safety concerns. The author concluded that the device could achieve significant results in fewer and shorter sessions, and that postpartum reshaping via circumferential reduction and skin laxity improvement can be effectively and safely achieved using a combination of RF energy, IR light and mechanical manipulation of the tissues.

Yu et al. [20] conducted a study in 19 patients with facial skin laxity and periorbital rhytides using the ReFirme ST Applicator (Syneron Medical Ltd., Yokneam). The study participants who had skin phototypes ranging from III to V received a total of 3 treatments performed in 3-week intervals to the whole face and neck regions. At 3 months after the last treatment, results showed that 89.5% of patients demonstrated moderate to significant subjective improvement in skin laxity of the cheek, jowl, periorbital area and upper neck, with a high overall satisfaction rating. In addition, mild improvement in skin laxity was observed over the mid- and lower face. The authors concluded that the combination of broadband IR light and bipolar RF can result in mild improvement of facial skin laxity in Asians with a high patient satisfaction, without any serious side effects from treatment.

Acne Vulgaris and Acne Scars

RF-based devices and their combinations with laser and light sources are proving to have utility in the treatment of acne vulgaris and the cosmetic improvement of acne scar lesions. Used alone or in combination, numerous laser and light sources including blue-, red-, green- and yellow-light sources as well as intense pulsed light, photodynamic therapy, diode lasers and RF devices are being used.

Taub and Garretson [21] conducted a clinical trial evaluating the safety and efficacy of sublative fractional bipolar RF and bipolar RF combined with diode laser for the treatment of patients with both superficial and deep acne scars. The 20-patient study included acne scar patients with skin types II–V who received up to 5 treatments performed in 4-week intervals that were directed to at least two facial (forehead, perioral, cheeks) and/or neck areas with acne scars. In the study, a combination of bipolar RF and 915-nm diode laser energy (Matrix IR, Syneron Medical Ltd., Yokneam) was used, followed by sublative bipolar RF energy (Matrix RF SR, Syneron Medical Ltd.,

Yokneam). Both applicator handpieces derived energy from a common platform (eLaser, Syneron Medical Ltd., Yokneam). Results showed that the acne scar lesions improved significantly one month after 3 treatments, and improvement persisted for at least 12 weeks after the fifth treatment. The authors concluded that the combination of diode laser and bipolar RF energy devices in addition to fractionated sublative RF is a safe and effective combined modality for the treatment of patients with both superficial and deep acne scars.

Prieto et al. [22] conducted a study evaluating the safety and efficacy of a combination of pulsed light and RF energy in 32 patients with moderate acne. Study participants received twice weekly treatments for 4 weeks with the Aurora AC device (Syneron Medical Ltd., Yokneam). In the 25 patients who completed the study, results showed that the mean lesion count was reduced by 47% after 8 treatments. In addition, the investigators found that the percentage of follicles with perifolliculitis decreased from 58 to 33%, sebaceous gland areas decreased from 0.092 to 0.07 mm^2, and heat shock protein 70 and procollagen-1 expressions did not change. They concluded that the combination of optical and RF energies may be an alternative nonablative modality for the treatment of moderate acne, and suggested that the clinical improvement seen may be in part due to reductions in both perifollicular inflammation and sebaceous gland areas.

Hair Removal

Although different lasers and light sources are used in hair removal treatments, one of the central limiting factors of these devices is the potential development of postinflammatory hyperpigmentation (PIH), particularly in patients with darker skin types. RF-based devices in combination with light energy have also been shown to be effective in hair removal. One of the main advantages of combining the RF and light sources is that lower optical fluences can be used due to the addition of RF energy, thereby significantly reducing the risk of PIH and resulting in safer treatments.

Yaghmai et al. [23] performed a multicenter clinical trial evaluating the safety and efficacy of the Aurora DS (Syneron Medical Ltd., Yokneam) in hair removal treatments. Of the 87 patients who were enrolled in the trial, 69 patients could complete the study. Study participants had Fitzpatrick skin types ranging from IV to VI, and received a singular treatment pass with the device on a specific body site. Results showed that the hair counts were reduced from baseline after one treatment by an average of 46%, and individual patient data showed that the percentage in hair count reduction achieved ranged from 0 to 100%, with 43% of patients having a 50% or greater reduction. The investigators concluded that the combination of optical energy and RF when delivered simultaneously can achieve effective hair reduction with the use of less optical energy, allowing for safer treatments of all skin types.

Sadick and Laughlin [24] also conducted a clinical trial evaluating the safety and efficacy of the combination of bipolar RF and intense pulsed light for hair removal in 36 female patients. Study participants with white and blond hair and skin phototypes ranging from I to V were treated for hair removal on the chin and upper lip, receiving a total of 4 treatment sessions over 9–12 months. The level of RF energy used was 20 J/cm^3, while fluences varied from 24–30 J/cm^2. Results showed an average hair removal of 48% at month 18 (6 months after the final treatment session). The investigators noted that a slightly higher photoepilatory efficiency was seen for blond hair (52%) compared to white hair (44%). They concluded that the combined RF and optical energy technology used can result in an effective photoepilation of blond and white hair phenotypes. This study has sparked debate, as it is generally held that hair which has low melanin content does not respond effectively to standard light-based treatment.

Vascular Treatments

Lapidoth et al. [25] conducted a study in 6 patients with lymphangioma circumscriptum, a benign acquired vascular lesion, using the Polaris LV system (Syneron Medical Ltd., Yokneam), a device that combines RF current and a 900-nm diode laser, evaluating safety and efficacy. The purpose of the study was to determine the safety and efficacy of the device for the treatment of lymphangioma circumscriptum, a lesion that due to its persistent nature and large variations in size, depth and anatomical location can pose great therapeutic challenges. Study participants (Fitzpatrick skin types II–III) received 1–3 sessions of simultaneous RF energy at 60–80 J/cm^3 and optical diode laser energy with fluences ranging from 90 to 100 J/cm^2. Results of the study were rated as 'excellent' in 4 patients and 'good' in 2 patients. Side effects included transient edema, erythema and pain in all patients, and ulcers and scarring in 2 patients. The authors concluded that the combination of laser light and RF energy was effective and relatively safe for the treatment of this notoriously challenging to treat vascular lesion. They found that the treatment modality can provide additional heating of the blood vessels without increasing the intensity of the laser, allowing the clinician to treat the clear lymphatic component, which lacks a specific chromophore.

Lapidoth et al. [26] also used the Polaris LV in another study, evaluating the safety and efficacy of the device for the treatment of facial venous malformations. The clinical trial included 14 patients with Fitzpatrick skin types II–IV who were treated for different types of facial venous malformations, both focal and multifocal, all measuring less than 100 cm^2, and located on various anatomic sites including the upper and lower lip, tongue, buccal area, cheek and forehead. Patients received a total of 1–3 treatment sessions with the device, and lesion clearance was evaluated by digital photographs taken at baseline and at 1 and 2 months after the last treatment. Results showed that 13 patients had a complete response (rated 'excellent') and 1 had a partial response (rated 'good'). Side effects of treatment included transient edema, erythema and pain in all patients, with permanent scarring seen in only one patient. The authors found the combination of RF and 900-nm diode laser energies to be both safe and effective for the treatment of facial venous malformations.

Conclusion

Aesthetic devices based on RF and in combination with novel laser and light energy sources and mechanical tissue manipulation technologies can offer patients a noninvasive therapeutic option to traditional and surgical approaches. The use of these devices has also been shown to treat a wide variety of indications that span both the dermatologic and aesthetic fields of medicine. The popularity of these noninvasive devices has significantly grown in recent years as they have been shown to achieve cosmetic outcomes that in many cases can rival those once only achievable with cosmetic surgery and due to their noninvasive nature, typically result in minimal side effects and downtime, particularly when compared to more traditional surgical techniques.

Studies have shown that combination RF treatments can achieve a synergistic effect in targeted tissues, resulting in the aesthetic improvement of a myriad of cosmetic indications. The therapeutic potential for RF-based devices used alone or in combination with other energy-based sources such as laser and light as well as mechanical tissue manipulation with pulsed vacuum and massage remains unknown as new treatment indications arise through continued research.

References

1 Elsaie ML: Cutaneous remodeling and photorejuvenation using radiofrequency devices. Indian J Dermatol 2009;54:201–205.
2 Alster TS, Lupton JR: Nonablative cutaneous remodeling using radiofrequency devices. Clin Dermatol 2007;25:487–491.
3 El-Domyati M, El-Ammawi TS, Medhat W, Moawad O, Mahoney MG, Uitto J: Electro-optical synergy technique: a new and effective nonablative approach to skin aging. J Clin Aesthet Dermatol 2010;3:22–30.
4 Sadick NS, Alexiades-Armenakas M, Bitter P Jr, Hruza G, Mulholland RS: Enhanced full-face skin rejuvenation using synchronous intense pulsed optical and conducted bipolar radiofrequency energy (elōs): introducing selective radiophotothermolysis. J Drugs Dermatol 2005;4:181–186.
5 Doshi SN, Alster RS: Combination radiofrequency and diode laser for treatment of facial rhytides and skin laxity. J Cosmet Laser Ther 2005;7:11–15.
6 Hammes S, Greve B, Raulin C: Electro-optical synergy (ELOS) technology for non-ablative skin rejuvenation: a preliminary prospective study. J Eur Acad Dermatol Venereol 2006;20:1070–1075.
7 Sadick NS, Trelles MA: Nonablative wrinkle treatment of the face and neck using a combined diode laser and radiofrequency technology. Dermatol Surg 2005;31:1695–1699.
8 Ryu HW, Kim SA, Jung HR, Ryoo YW, Lee KS, Cho JW: Clinical improvement of striae distensae in Korean patients using a combination of fractionated microneedle radiofrequency and fractional carbon dioxide laser. Dermatol Surg 2013;39:1452–1458.
9 Issa MC, de Britto Pereira Kassuga LE, Chevrand NS, do Nascimento Barbosa L, Luiz RR, Pantaleão L, Vilar EG, Rochael MC: Transepidermal retinoic acid delivery using ablative fractional radiofrequency associated with acoustic pressure ultrasound for stretch marks treatment. Lasers Surg Med 2013;45:81–88.

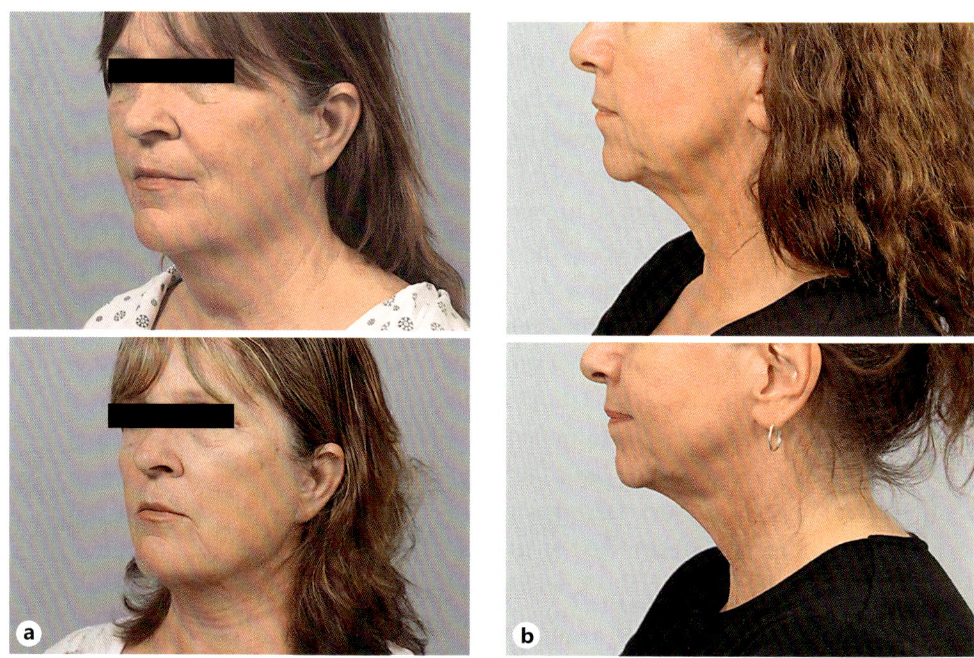

Fig. 3. a Dramatic improvement: 68-year-old before and 6 weeks after RF-assisted neck lift. **b** Modest result: 54-year-old with low hyoid bone has less dramatic outcome with the same procedure.

Prolonged temporary deformities are rarely seen when noninvasive RF modalities are used. Generally, these include small burns, focal depressions, or PIH type concerns. If these do not resolve within 6–8 weeks, further treatment options should be offered to the patient.

The lack of expected clinical response within a 6- to 8-week period can be troublesome to the patient and provider. Patients who have high expectations are quick to complain and slow to be reassured. The best prevention in this case is preoperative education, and a clear written explanation that results vary from patient to patient (fig. 3). However, if a conservative 'undertreatment' was performed, either due to inexperience on the part of the practitioner, or an unwillingness to fully utilize a device with consumables, the practitioner may be obligated to offer a second treatment at little or no charge to the patient. Both of these eventualities might be considered an error in judgment. A keen sense of this is needed by the aesthetic practitioner. The calculated balance of a thorough and full treatment session against the risk of a subsequent complication is sometimes difficult, even for the experienced and savvy physician. When practice economics enter the picture, decision making can become even more clouded. Priorities in decision making should include the best safe result for the patient as tops, with the concern regarding the practice's balance sheet as secondary. The latter should not be totally discounted, however, as the cost of consumables associated with the use of some devices can easily obliterate any profit for the practice.

Prevention of Minor Complications

The avoidance of sequelae or complications requires good planning prior to the performance of any cosmetic procedure. The first consideration should always be: should I treat this patient? Because many noninvasive treatments are considered low risk, and are often done by practice extenders, direct physician supervision may not be routinely done. The fact that a practitioner owns a device does not mean that it should be used on every patient. Meeting expectations without creating a new problem is a good basic premise. The responsible practitioner will discuss all possibilities without overselling the treatment. In order to reduce patient dissatisfaction, it is good medical practice for the supervising physician to be present during the initial consultation and treatment if practice extenders are utilized.

According to Kreindel [17], the one setting that most directly affects tissue response is the temperature achieved during treatment. Safety can be maximized by using the lowest effective temperature setting that will achieve the proper outcome. Other parameters that affect safety include pulse and treatment duration, power of each pulse, whether the energy is pulsed or continuous, and the delivery type of RF energy, such as monopolar or bipolar. Pulse stacking is contraindicated as this can cause focal fat atrophy, sometimes noted to appear in the shape of the treatment tip. When using RF fractional devices, conservative overlap of pulses will prevent skip areas requiring subsequent corrective treatment. Similar attention to detail should be used with 'stamping' devices targeted towards tissue tightening and skin rejuvenation. The use of vacuum-assisted tissue-tightening or fat-reducing devices should also be done with a slight overlap.

Prevention of burns can be achieved by reducing energy settings [18] and by monitoring the skin temperature with an infrared thermometer or 'temperature gun', while limiting the maximum skin temperature to 40°C or less. Assuring that full skin contact is achieved with every pulse when using bipolar or multipolar devices can reduce the risk of arc burns. Moving from one treatment region to the next can reduce unexpected build-up of thermal energy. While disappointment in outcome is a risk of undertreatment, overtreatment can cause permanent deformities that cannot be repaired with an extra treatment.

Minimally Invasive Treatment with Radiofrequency-Assisted Devices: Minor Complications

Complications directly caused by the minimally invasive RF device differ significantly from those that can follow treatment with noninvasive devices. The use of RF energy in the subcutaneous plane generates another level of risk. While end hits and superficial burns generally heal without long-term sequelae, a raft of associated concerns accompanies treatments located in the hypodermis. Possible minor complications that can occur with minimally invasive RF-assisted treatments include: first or second degree burns, small area; small seromas, <100 ml; small hematomas, <100 ml; subcutaneous nodularity; skin contour irregularities; residual skin laxity, minor; burns, including periportal erosions; mild hyperpigmentation requiring no intervention; unsatisfactory scarring; prolonged discomfort or neuropathy; less than hoped for result.

Prevention and Treatment

While minor complications are technically defined as those needing no further intervention, frequently the severity and duration of these problems can be minimized with early palliative measures. Because the possibility of thermal damage is always present with heat-mediated treatments, the immediate availability of cold compresses is a vital element during the procedure. Application of a cold damp compress is indicated when focal erythema, an arc burn, or blister is noted during treatment. If, during treatment, the practitioner feels a 'hot spot' with his or her hand, serial application of a cool compress can significantly reduce the risk of a burn, focal depression, or seroma. If the treatment physician can choose a device with an external or internal skin temperature monitor, this can also reduce the risk of burns. It is important to understand the isotherm of the device that you are using. Small-diameter monopolar devices are likely to reach a focal very high temperature [12], especially if the treatment region is small. In a series of 10 patients, infrared monitoring of sequential treatment regions showed that even after this type of device was removed, the tissue temperature continued to rise for an average of 2 min. Once a 'heat sink' is created, the temperature can rise rapidly, causing the emergence of a late unsuspected burn. In order to prevent this, moving to another area – whether deep, superficial, adjacent, or distant – is advisable. It is always better to let the treatment region cool down, then repeat heating, than to create a pool of superheated fat which can cause a burn, oil cyst, or seroma. Avoidance of spending too much time in the deep abdominal subcutaneous fat will reduce the risk of a seroma in that region.

While small hematomas can be aspirated, many resolve without treatment. Risk reduction includes having the patient abstain from taking any aspirin, NSAID, warfarin, or other 'blood thinners' (including certain vitamins and herbal supplements) for at least 9 days prior to treatment. Subcutaneous nodularity was noted in 3/50 patients treated with subcutaneous RF-assisted tissue tightening for breast lifting [19]. One patient had spontaneous resolution and 2 were treated with intralesional collagenase injections with complete resolution. When using an RF device for the purpose of tissue tightening, the fibroseptal network is targeted [20]. Coagulation or 'drawing together' of the fibroseptal network is expected, and can cause some temporary stiffness or fibrosis following treatment.

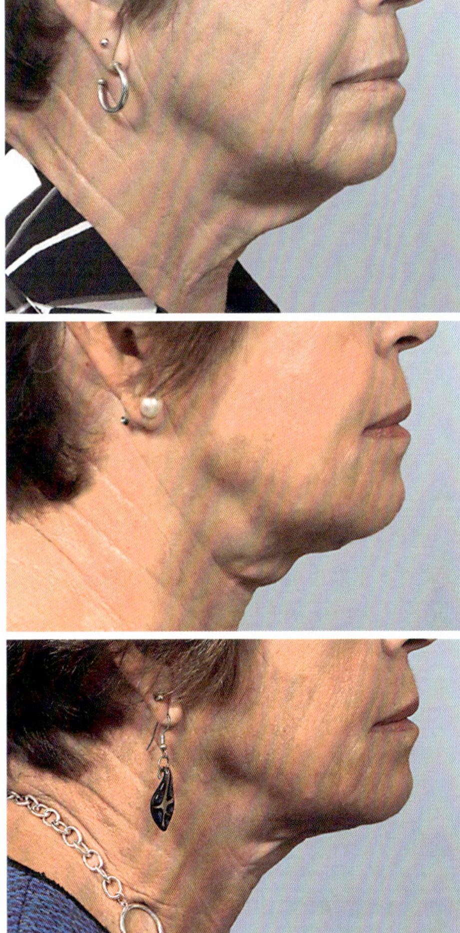

Fig. 4. Prolonged temporary deformity following RF-assisted necklift. This 60-year-old before treatment (above). The middle image shows an early result 2 months postoperatively. The patient was unhappy with the lumpiness and stiff tissue at the cervicomental junction. The bottom image shows her appearance at 4 months postoperatively, with no further treatment.

Skin contour irregularities following treatment in the subcutaneous plane are not uncommon if one takes a critical look. Rosato [pers. comment] noted that approximately 40% of his post-liposuction patients had a visible contour irregularity, but Pitman and Temourian [21] stated that only 10% require surgical revision. Figure 4 shows a patient who did not wear the recommended postoperative compression garment. At 3 months after operation, she still had significant lumpiness. However, at 6 months, the contour deformity had largely resolved without surgical intervention.

Residual skin laxity, if recognized early in the posttreatment course, can be minimized by applying localized compression, for example with silicone-gel bandages. In patients with very thin skin, self-correction is less likely. With these patients, it is important to avoid overresection of fat in the superficial regions while performing lipoaspiration. Otherwise, a condition colloquially known as 'chicken skin' can occur. This type of crepe skin appearance is very difficult to improve, although mild to moderate improvement can be seen with laser or RF resurfacing.

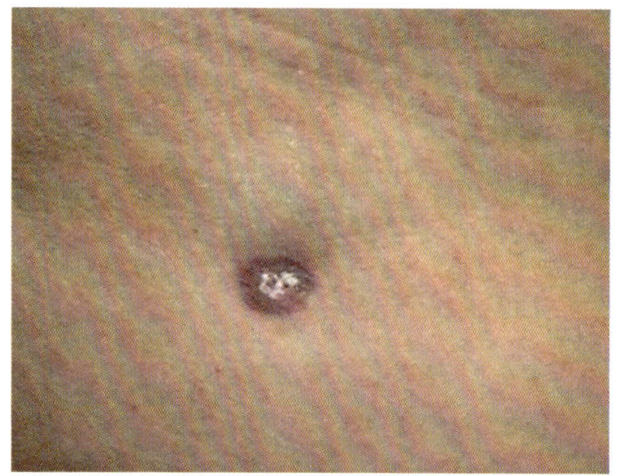

Fig. 5. Periportal burn 4 weeks after RF-assisted liposuction.

Periportal burns (fig. 5) or erosions can take a long time to improve. If the patient becomes impatient, further treatment with intense pulsed light (IPL) or RF resurfacing can improve prolonged erythema in these areas. Unsatisfactory scarring is frequently due, in the early postoperative phase, to prolonged erythema. The same treatment options can be used to reduce the red or pink character of the access ports.

Prolonged discomfort or neuropathy is an unusual but perplexing complaint. On occasion, it can be due to fibrosis in the location of sensory nerves. Occasionally, a collagenase injection can help, and gentle local massage can be used to reduce symptoms. As time progresses, most of these 'healing pains' diminish. Mild anti-inflammatory drugs can be used to treat more severe symptoms.

Major Complications Caused by Radiofrequency-Assisted Aesthetic Treatments

Possible major complications that can occur with RF-assisted aesthetic treatments include: large second or any third degree burn; seroma >100 ml; hematoma >100 ml; contour irregularity requiring treatment such as subcision, collagenase injection, further liposuction, or fat grafting for correction; postinflammatory hyperpigmentation requiring prolonged application of topicals or IPL; unsatisfactory scarring requiring revision; asymmetry requiring further treatment; focal depression; oil cyst; residual skin laxity or textural abnormality.

Treatment

While small second-degree burns rarely require surgical intervention, burns that develop a component of full thickness skin loss almost always do. Direct excision with a precise layered closure is the best option, in most cases (fig. 6). If the third-degree burn is small or punctate, occasionally excision with a circular dermal punch can be effective. Centripetal healing forces can reduce the scar to a minimal size.

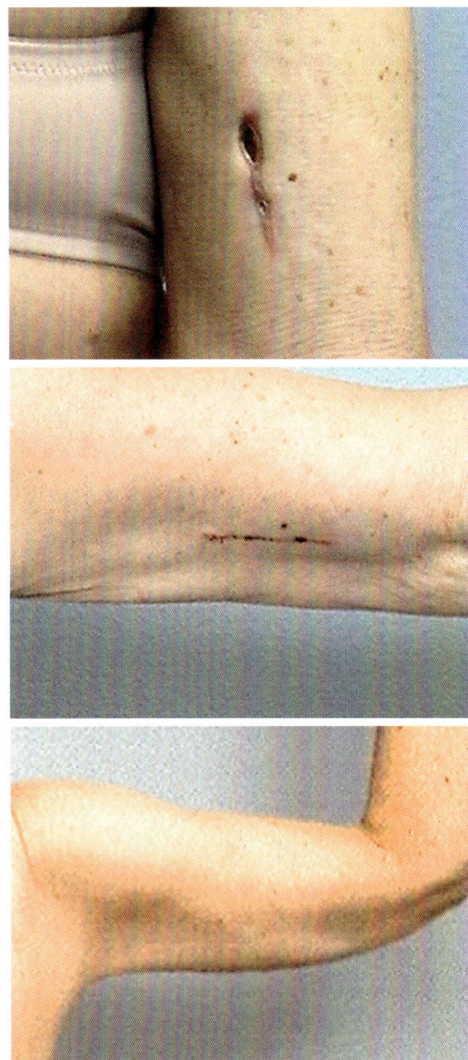

Fig. 6. Full-thickness burn following RF-assisted liposuction. Middle image shows immediate appearance following excision and layered closure. Bottom image shows a barely discernible scar.

On rare occasions, excessive thermal damage can be beneficial to the long-term result, although intentionally doing this is not at all advised (fig. 7). The patient was treated with a monopolar device, and had very thin skin. Despite controlled observation of an external skin temperature never exceeding 40°C during the procedure, she sustained a full-thickness burn that appeared on postoperative day 3. She was seen daily for a week, then weekly for 6 weeks. At that point, the area became quite small and hard to see. At 4 months, she had a remarkable result due to the large amount of thermal effect. While she was grateful for the outcome, the recovery period was filled with anxiety on her part and that of the treating physician. The lesson learned in this case was that clinical judgment and conservatism in treatment overrule thermometers or other device indicators.

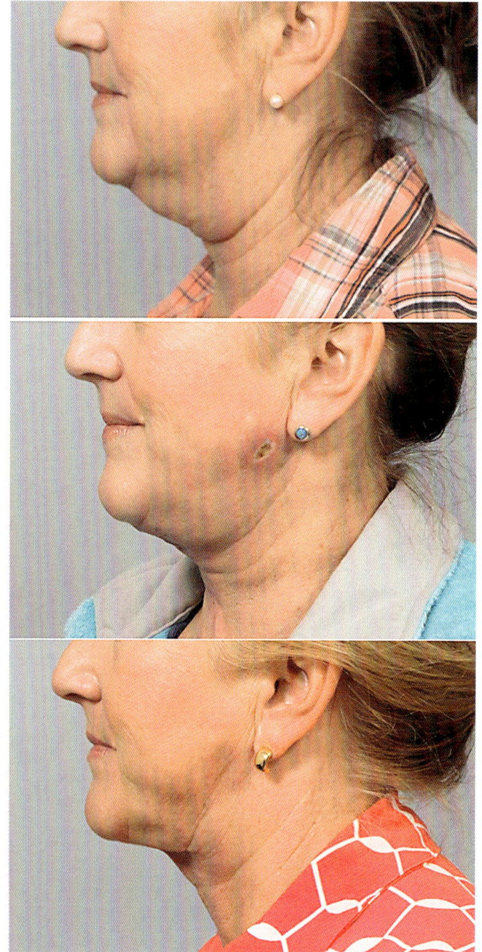

Fig. 7. 55-year-old with severe neck laxity, above. The patient underwent a monopolar RF-assisted neck lift with no skin excision. No blisters were noted immediately after operation. Ten days later, she presented with a small unexpected full-thickness burn. The degree of soft tissue tightening was remarkable, and the burn healed well without further treatment.

Large seromas and hematomas rarely heal without surgical intervention. While early hematoma evacuation is fairly straightforward, with seromas, full excision of the seroma 'lining' is needed to correct that problem completely. The use of 'quilting' sutures to fix the overlying tissue to the fascial base can reduce the risk of recurrence. Adjunctive use of a compression garment and strong activity restrictions can help, and should also be routinely used in the early perioperative period as a preventative measure.

Revisions for severe contour irregularities usually require some type of release of the depression, which can be done using a 'Rigottomy' technique [22], traditional subcision [23], laser-assisted release [24] or a 'pickle fork' device. Frequently, the depression also requires some type of fat transfer or filler to correct the problem adequately, especially if the depression is focal (fig. 8). Reduction of residual fatty protrusions can be accomplished by using controlled liposuction, or laser lipolysis [25] to 'melt' fat without aspiration, or, in carefully selected patients, injection lipolysis. Some oil cysts

Duncan

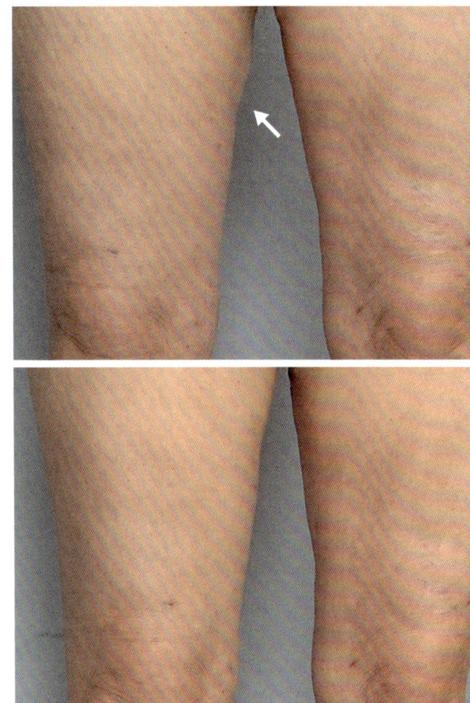

Fig. 8. This 61-year-old retiring schoolteacher had RF-assisted liposuction of her inner thighs. She had a depression in the right inner thigh that was not present preoperatively. Correction with subcision and fat grafting gave her an improvement in contour.

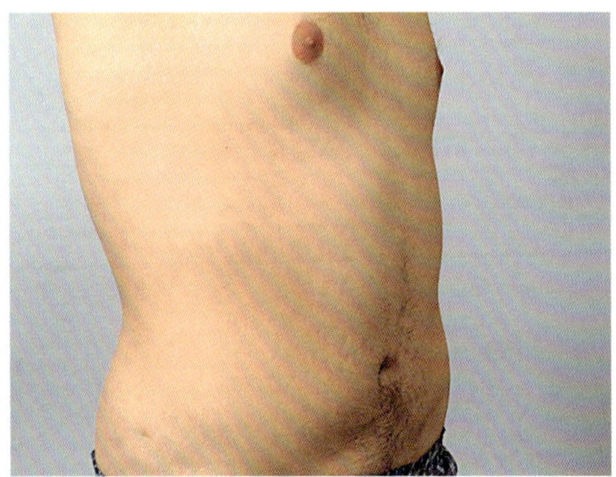

Fig. 9. This 21-year-old man underwent RF-assisted liposuction for lower abdominal and flank adiposity. He presented at one year after treatment with an oil cyst that was treated with aspiration.

resolve with simple aspiration, while others persist and may require excision in order to fully correct the deformity (fig. 9).

In patients with residual skin laxity following treatment with liposuction enhanced with RF-assisted tissue tightening, fractional laser or fractional RF resurfacing can improve minor deformities. In patients requiring more than 10% surface area reduction, direct skin excision is frequently the only realistic revisional option (fig. 10).

Periportal burns causing persistent erythema can be the biggest complaint that RF-assisted liposuction patients vocalize. While the redness fades with time, some patients

Fig. 10. This 43-year-old had a combination of upper arm adiposity and skin laxity. Liposuction plus brachioplasty was recommended. She declined this treatment as she did not want scars. This patient was advised that even with RF-assisted liposuction, she would have some residual skin excess. Her photo 2 months after treatment showed the expected residual excess skin, which was excised with a mini-brachioplasty.

demand an accelerated improvement. Treatment with IPL of vascular lasers can help. Because of the coagulative nature of RF energy, newer RF fractional devices can also help with 'red' reduction. Occasionally, a scar revision is indicated. However, because the second scar is inevitably longer than the first, and the cycle of inflammatory erythema begins anew, scar revision is reserved for severe cases.

Ancillary Complications Associated with Radiofrequency-Assisted Cosmetic Treatments

While these complications can be serious, ancillary consequences of treatment with RF-assisted devices are not directly caused by the RF component. Rather, they are known complications due to simultaneously performed procedures such as liposuction.

While life-threatening complications are generally rare, they can occur [26]. Surgeons who utilize traditional liposuction or RF-assisted liposuction can cause a perforated viscus or another intra-abdominal injury, including major vascular perforations. Lidocaine toxicity can be fatal. The creation of a fat embolus is possible, as disrupted or liquefied fat near a blood vessel can enter that vascular channel if a perforation occurs. While not directly caused by the use of RF energy, thromboemboli can occur during prolonged combined or prolonged procedures, especially those with an orthopedic, abdominoplasty, or gynecologic component. A risk assessment evaluation should be completed for all patients undergoing RF-assisted liposuction or extensive RF-assisted tissue tightening. Congestive heart failure or renal insufficiency can occur in patients that are predisposed in cases where tumescent infusion is poorly managed. Though recent literature shows that lidocaine toxicity is rare [27], some patients may experience perioral tingling or other manifestations when large areas are treated. Because many RF-assisted procedures are not considered 'invasive' or risky, sometimes a less than thorough history and physical exam can be performed. A good safety strategy is to treat all patients as if they were having major surgery at the time of initial evaluation.

Conclusions

RF-based aesthetic devices are available for many purposes. They range from noninvasive to minimally invasive, though RF-assisted liposuction can be considered a major procedure. The learning curve can be short, or quite long. Because heat-mediated tissue response has been well studied [28–30], consistent and predictable results are becoming the norm with many approved devices. Associated risks should be considered a limiting factor by the device operator. Canizarro's rule [Canizarro, pers. comment] is a good guideline for the treating physician: 'if you cannot adequately treat the possible complications, don't do the procedure'. While this admonition was generally directed towards overeager surgical residents wanting to perform cholecystectomies, the rule is even more appropriate today, when nonspecialty physicians are branching out into the cosmetic field. A good basic knowledge of the way RF energy works accompanied by a careful patient assessment and choice of the proper procedure will go a long way towards reducing risks of treatment with RF-assisted aesthetic devices.

References

1 Krueger N, Sadick NS: New-generation radiofrequency technology. Cutis 2013;91:39–46.
2 Paul M, Blugerman G, Kreindel M, Mulholland RS: Three-dimensional radiofrequency tissue tightening: a proposed mechanism and applications for body contouring. Aesthetic Plast Surg 2011;35:87–95.
3 Taub AF, Garretson CB: Treatment of acne scars of skin types II to V by sublative fractional bipolar radiofrequency and bipolar radiofrequency combined with diode laser. J Clin Aesthet Dermatol 2011; 4:18–27.
4 Blugerman G, Schavelzon D, Paul MD: A clinical and histological study of radiofrequency-assisted liposuction (RFAL) for aesthetic reshaping and skin tightening. Aesthet Plast Surg 2012;36:62–71.

5 Ahn DH, Mulholland RS, Duncan DI, Paul MD: Non-excisional face and neck tightening using a novel subdermal radiofrequency-assisted thermocoagulative device. J Cosmet Dermatol App 2011, Epub ahead of print.

6 Paul MD, Mulholland RS: A new approach for adipose tissue treatment and body contouring using radiofrequency-assisted liposuction. Aesthet Plast Surg 2009;33:687–694.

7 Sadick NS, Sato M, Palmisano D, Frank I, Cohen H, Harth Y: In vivo animal histology and clinical evaluation of multisource fractional radiofrequency skin resurfacing (FSR) applicator. J Cosmet Laser Ther 2011;13:204–209.

8 Theodorou SJ, Paresi RJ, Chia CT: Radiofrequency-assisted liposuction device for body contouring: 97 patients under local anesthesia. Aesthetic Plast Surg 2012;36:767–779.

9 Atiyeh BS, Dibo SA: Nonsurgical nonablative treatment of aging skin: radiofrequency technologies between aggressive marketing and evidence-based efficiency. Aesthet Plast Surg 2009;33:283–294.

10 Mulholland RS: Noninvasive body contouring with radiofrequency, ultrasound, cryolipolysis, and low level laser therapy. Colin Plast Surg 2011;38:503–520.

11 Weiss RA, Weiss MA, Munavalli G, Beasley KL: Monopolar radiofrequency facial tightening: a retrospective analysis of efficacy and safety in over 600 treatments. J Drugs Dermatol 2006;5:707–712.

12 Duncan D, McBurney D, Kinney B: Focal nerve lesions with a monopolar frequency device for cosmetic enhancement; in Di Giuseppe A, Shiffman MA (eds): New Frontiers in Cosmetic and Plastic Surgery. Oak Park, Bentham, in publication.

13 Rigopoulos D, Gregoriou S, Katsambas A: Hyperpigmentation and melasma. J Cosmet Dermatol 2007;6:195–202.

14 Sadick N: Bipolar radiofrequency for facial rejuvenation. Facial Plast Surg Clin North Am 2007;15:161–167.

15 Davis EC, Callender VD: Postinflammatory hyperpigmentation: a review of the epidemiology, clinical features, and treatment options in skin of color. J Clin Aesthet Dermatol 2010;3:20–31.

16 Hatipoglu N, Hatipolglu H: Combination antifungal therapy for invasive fungal infections in children and adults. Expert Rev Anti Infect Ther 2013;11:523–535. Review

17 Kreindel M: Radiofrequency basics; in IMCAS Asia, Hong Kong, October 2012.

18 Sadick N: Tissue tightening technologies: fact or fiction. Aesthet Surg J 2008;28:180–188.

19 Duncan D: 'Scarless' mastopexy: using heat mediated tissue tightening to lift the breast; in Di Giuseppe A, Shiffman MA (eds): New Frontiers in Cosmetic and Plastic Surgery. Oak Park, Bentham, in publication.

20 Mulholland RS, Paul MD, Chalfoun C: Noninvasive body contouring with radiofrequency, ultrasound, cryolipolysis, and low-level laser therapy. Clin Plast Surg 2011;38:503–520.

21 Pitman G, Temourian B: Suction lipectomy: complications and results by survey. Plast Recon Surg 1985;76:65–69.

22 http://www.intechopen.com/download/pdf/27951 (accessed 04/12/13).

23 Hexsel D, Dal' Forno T, Soirefmann M, Hexsel CL: Reduction of cellulite with Subcision®; in Murad A, Pongprutthipan M: Body Rejuvenation. Berlin, Springer, 2010, pp 167–171.

24 DiBernardo BE: Randomized, blinded split abdomen study evaluating skin shrinkage and skin tightening in laser-assisted liposuction versus liposuction control. Aesthet Surg J 2010;30:593–602.

25 Tierney EP, Kouba DJ, Hanke CW: Safety of tumescent and laser-assisted liposuction: review of the literature. J Drugs Dermatol 2011;10:1363–1369.

26 Rao RB, Ely SF, Hoffman RS: Deaths related to liposuction. N Engl J Med 1999;340:1471–1475.

27 Klein JA: The tumescent technique. Anesthesia and modified liposuction technique. Dermatol Clin 1990;8:425–437.

28 Vangsness CT Jr, Mitchell W III, Nimni M, Erlich M, Saadat V, Schmotzer H: Collagen shortening. An experimental approach with heat. Clin Orthop Relat Res 1997;337:267–271.

29 Lin SJ, Hsiao CY, Sun Y, Lo W, et al: Monitoring the thermally induced structural transitions of collagen by use of second-harmonic generation microscopy. Opt Lett 2005;30:622–624.

30 Hayashi K, Thabit G III, Massa KL, et al: The effect of thermal heating on the length and histologic properties of the glenohumeral joint capsule. Am J Sports Med 1997;25:107–112.

Diane Irvine Duncan, MD, FACS
Plastic Surgical Associates of Fort Collins
1701 East Prospect Road
Fort Collins, CO 80525 (USA)
E-Mail dduncanmd@aol.com

Author Index

Subject Index